SpringerBriefs in Applied Sciences and Technology

Manufacturing and Surface Engineering

Series editor

Joao Paulo Davim, Aveiro, Portugal

For further volumes:
http://www.springer.com/series/10623

Harun Pirim · Umar Al-Turki
Bekir Sami Yilbas

Supply Chain Management and Optimization in Manufacturing

Springer

Harun Pirim
Umar Al-Turki
King Fahd University of Petroleum
and Minerals
Dhahran
Saudi Arabia

Bekir Sami Yilbas
Department of Mechanical Engineering
King Fahd University of Petroleum
and Minerals
Dhahran
Saudi Arabia

ISSN 2191-530X   ISSN 2191-5318 (electronic)
ISBN 978-3-319-08182-3   ISBN 978-3-319-08183-0 (eBook)
DOI 10.1007/978-3-319-08183-0
Springer Cham Heidelberg New York Dordrecht London

Library of Congress Control Number: 2014942272

Printed on acid-free paper

Springer is part of Springer Science+Business Media (www.springer.com)

Preface

We are living in such a connected century that networks (e.g., social networks) are attracting more and more attentions of people every day. Twitter, Facebook, and LinkedIn are among the most celebrated companies.

We can easily manifest that the success comes from understanding the importance of collaboration and network science. In the manufacturing context, even if manufacturing itself is very efficient a company can hardly compete with others through more expenditure in it. Rather, one should look into procurement, distribution channels, facility, and inventory decisions as a whole in order to compete to satisfy the high quality needs of customers at a reasonable (i.e., minimal) cost. Supply chain management has remained as one of the hottest topics for decades for this case. However, supply chain design and improvement in any area of supply chain requires integration and engagement to justify the fact that the strength of a chain is due to its weakest link. Then, as one would agree, one's genius can hardly beat a collective genius.

This book introduces state-of-the-art supply chain management topics keeping it brief enough for novice readers and deep enough for researchers in the field. The book adopts both management and optimization paradigms.

Management topics include strategic level organization and planning-related subjects. Optimization topics review important optimization models for supply chain-wide location, production, and transportation problems. Solution procedures are discussed as well. Illustrative examples are provided. Each chapter ends with remarks providing the core ideas of the chapters.

The book starts with an *Introduction* and the second chapter deals with *Supply Chain Management*. This chapter discusses key decisions in supply chain management and considers planning operations for it. The third chapter introduces *Scheduling Models in Supply Chain*. The last chapter is *Optimization in Supply Chain*. Optimization problems and models reviewed are classified under transportation and facility location.

Acknowledgments

We acknowledge the role of King Fahd University of Petroleum and Minerals in extending strong support from beginning to end facilitating every means during the preparation of the book. The authors wish to thank the colleagues who contributed to the work presented in the book through previous cooperation of the authors. In particular, thanks to all our graduate students.

Contents

Contents

Chapter 1
Introduction

Abstract Supply chain is a complex network with multiple layers such as suppliers, manufacturers, warehouses, distributors, retailers, and customers. Supply chain management and optimization requires conflicting decisions, coordination and integration. Response time, product variability may affect supply chain design decisions. Problems associated with supply chain may be approached by mathematical modeling and optimization. The challenge through the supply chain is being responsive to customers with cost efficiency. Manufacturing processes affect the efficiency of the supply chain. This chapter presents a concise introduction to main concepts and topics of supply chain management and optimization.

Keywords Supply chain management · Design decisions · Efficiency · Response time

Manufacturing is defined in an industrial context where resources such as labor, machines, raw materials etc. are utilized to produce a physical output, product. A supply chain in manufacturing includes suppliers, manufacturing plants, warehouses, and retailers as stakeholders. A supply chain also comprises activities within and between stakeholders that integrate them. Chopra and Meindl [3] define supply chain as all stages involved directly or indirectly satisfying customer demands. They mention that in this regard a supply chain does not only include the manufacturer and suppliers but also includes transporters, warehouses, retailers, and customers. Many firms of the supply chain manufacture products adding value to the chain converting raw materials to final products [2].

"Supply chain management" term dates back to 1980s and it can be defined as the coordination of location, manufacturing, inventory, and transportation among the stakeholders of a supply chain to achieve the best mix of responsiveness and efficiency for the market served [5]. Optimization is necessary in supply chain management. All stages of a supply chain can be regarded as an optimization problem. Minimizing the total cost of transportation while satisfying customer needs, minimizing inventory holding costs throughout the supply chain while fulfilling the demands of plants or end customers, deciding on the best facility

H. Pirim et al., *Supply Chain Management and Optimization in Manufacturing*,
SpringerBriefs in Manufacturing and Surface Engineering,
DOI: 10.1007/978-3-319-08183-0_1, © The Author(s) 2014

location minimizing the distribution costs are some examples. Supply chain is so inherent in our life that we are at the center of producing or demanding something personally.

Main supply chain decisions are about facility location, production, inventory, and transportation. Supply chain decisions may be strategic, tactical, and operational. Decisions are triggered by both customer requirements and efficient supply chain operations. These decisions may conflict with each other. For instance, customer satisfaction and mass production to decrease the manufacturing costs result in high inventory levels. So, a production decision has a conflict with an inventory decision. Conflicting decisions require coordination and integration through supply chain to optimize the processes in it globally. One aspect of coordination and integration is partnership. The partnership (volume and complexity of manufacturing may prompt it) is justified if the manufacturers, together yield better results than before partnership [6]. Narayan and Raman [9] studying 50 supply networks find that companies that look out for their own interests ignoring their network partners have poor supply chain performance. They observe that a supply chain works effectively if the risks, costs, and rewards of executing operations are distributed fairly across the supply chain network. Otherwise, the supply chain will suffer with excess inventory, stock-outs, wrong demand forecasts, futile sales efforts, and poor customer service. To achieve overall supply chain efficiency, manufacturers may postpone their schedules with a per-unit-cost sacrifice [2]. Authors suggest that inventory availability, speed and consistency of delivery (operational performance), and efficient operations are the elements of a logistically sophisticated firm which is an ideal supply chain partner.

Supply chain design decisions are at strategic level. Determining the number, capacity and locations of plants, warehouses with the minimum cost, deciding on the flow of goods, funds, information and services, matching distribution centers with customers are examples [8]. There is not a unique solution for supply chain design problems. Addressing supply chain problems requires constructing mathematical models, solving the models and analyzing the results to make operational decisions [12]. Optimal decision determines the best supply chain for a product to flow from supplier to customer. Experts suggest that 80 % of the supply chain cost is incurred with the location of the facilities and product flows between them [12]. As mentioned by the authors, here are some questions to help design a supply chain network:

- What should be the number of warehouses? Where should they be located? What should their capacity be? Which products should be distributed from them? How different customers should be served from them?
- What should be the number of manufacturing plants? Where should they be located? What should their capacity be? How many production lines should a plant have? What products should be made? Which warehouses should they serve?
- Which products should be manufactured internally? Which products should be outsourced? In case of outsource, which suppliers should be used?

- What is the trade-off between the number of facilities and the supply chain costs?
- What is the trade-off between the number of facilities and customer service level? What is the cost of improving service level?
- How is the supply chain network affected if the demand, labor cost, raw material prices change?
- Regarding the seasonality of the products, when is the best time to produce?
- If the demand increases, should the capacities of existing plants be expanded or new plants should be added? When should it expanded or added?
- How can overall supply chain cost be reduced?
- How frequently should a supply chain be reevaluated for efficiency?

A supply chain should be re-evaluated periodically for efficiency. It is important to note that supply chain term is used in a broader sense instead of logistics. Traditional logistic activities include purchasing, distribution, maintenance and inventory management while supply chain management also includes marketing, finance, and customer service and product development [5]. Another distinction is that logistics include all activities to send and obtain products/information between supply chain stake holders, compared to the supply chain providing a framework and standard for members to satisfy customer needs [2].

Five areas of a logistical work are interrelated: orders, inventory, transportation, material handling and packaging, and facility network [2]. Orders are from the customers. Orders include receipt, delivery, and invoice. Inventory is kept at minimum satisfying the desired customer service. Raw materials, work-in-process, and final product make up inventory. Transportation moves the inventory. Consistency in transportation is the most important factor. Consistent transportation has little or no variation in time for specific shipments. Safety stocks may be used to compensate inconsistency. Then transportation speed with reasonable cost counts. The longest distance with the largest shipment should be taken to reduce the transportation cost. Handling and packaging effects inventory, transportation status. Handling and packaging systems are costly, however they aid in delivery efficiency with product safety. Facility network includes the location of facilities. Manufacturing plants, warehouses, distribution centers, and retailers are some of the facilities. It is important to identify the optimum number and location of these facilities, stock levels, and customer assignments keeping total logistics cost at minimum.

A supply chain can be viewed as a complex network composed of multiple echelons. Suppliers (S), manufacturers (M), warehouses (W), distributors (D), retailers (R), customers (C) are some of the echelons. The interrelations among the echelons are shown in Fig. 1.1.

An interesting responsive supply chain approach is from Zara. Zara keeps nearly half of its production in-house. Rather than asking its manufacturers to maximize the output, Zara focuses on building extra capacity, instead of economies of scale, Zara manufactures and distributes in small amounts, managing all production, inventory, and transportation functions itself [4]. Authors report that

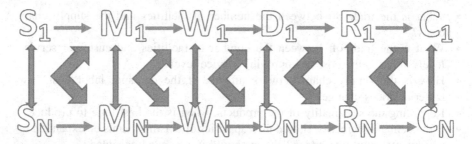

Fig. 1.1 Multi-echelon supply chain. All *arrows* indicate directions of possible commodity and information flow between echelons and N members of each echelon

Zara's responsive supply chain delivers 600 stores worldwide in 15 days. Zara offers new products with limited quantity. That is a supply chain model for products with short life-cycles.

In 1990s, an average time for a company to deliver a product from warehouse to customer would take around a month and even more when something goes wrong such as lost orders, out-of-stock case, and misdirected shipments [2]. Supply chain activities used to include order-to-delivery processes that include:

1. order initiation: transfer by telephone, fax, electronic data interchange, mail
2. order process: manual or computer systems, credit authorization, assignment to a warehouse
3. shipment [2].

Another supply chain example is from Toyota. It is appreciable that every activity, connection, and production flow is rigidly documented in a Toyota plant, while Toyota's operations are very flexible and responsive to customer demands [11]. Authors mention that Toyota's rigidity of the operations make the production flexible. The way workers work, interact with each other, learn to improve, and production lines are constructed, are the principles determining how Toyota deals with its operations as experiments and teaching scientific methods to its workers. Spear [10] states that many manufacturers adopt Toyota Production System successfully, while some firms don't have success stories since they only focus on specific TPS tools and practices without recognizing the underlying philosophy.

Today's industry challenge is keeping up with the speedy response to customer at a reasonable cost. The challenge draws attentions to breaking down the internal barriers and establishing effective cross-functional relationships [1]. However, Lee [7] mentions after spending 15 years studying more than 60 companies that only companies with agile, adaptable, and aligned supply chains can address the challenge.

Today's supply chain practices evolved by many years of experience from industrial revolution, that is the world is not characterized by scarcity but challenged with a variety of customer needs [2]. In other words, a customer's passive acceptance of a product shifted to active encounter in the design and delivery of a product [2]. A customer demands high quality at a low price. In manufacturing context, customers ask for increasing the functionality of a product integrating customer specified components. Information technologies brought a rapid global economy allowing response to customer in a very short time. Now, delivering on time the desired quantity of products with a guaranteed quality is an expectation rather than an exception. Such high level supply chain performance is achieved at a lower cost than the past [2].

Performance of a supply chain distribution network is evaluated through two dimensions. These are the customer needs and the cost of meeting customer needs. Therefore, a trade-off, between satisfying customer needs and keeping transportation cost minimum, is made in order to decide which distribution network is better.

There are many aspects which effects customer service. Some of them are directly influenced by the structure of the distribution network [3]: response time, product variety, product availability, customer experience, time to market, order visibility, and returnability. These factors, as explained below, affect supply chain network design decisions.

Response time is the amount of time it takes a customer to receive his/her order. Book stores such as Barnes and Noble respond to customers just in time whereas Amazon handles book orders in two days or more. However, Amazon has a variety of books more than any other book store. We can say that Amazon satisfies the variety of customer needs. Product variety is the number of different products or configurations that are offered by the distribution network.

Product availability is likelihood of having a product in stock when a customer order arrives. As the number of facilities increase, the amounts of inventories increase as well. That may result in product availability while increasing the inventory costs. We can say that transportation costs decrease to some extent when number of facilities increase.

Customer experience comprises ease and customization of orders, value gained during sales process.

Time to market is the time needed to bring a new product to market. For customers demanding especially new technologies, time to market is a critical issue.

Order visibility is the track of orders from placement to delivery. Order visibility is very important for convenience of a firm. It needs a good information infrastructure. DHL and similar firms have tracking options. Order visibility saves labor hours for a firm. Integration of supplier and manufacturer is an important issue in order visibility.

Returnability is the ease of sending back unsatisfactory merchandise and ability of a network to handle returns.

According to all of these factors, a firm decides what kind of distribution network design is suitable. Changing a distribution network design affects inventories, transportation, facilities and handling, information. Accurate and timely information is crucial in supply chain operations for customer's order monitoring, reducing excess inventory, need for labor, and aiding supply chain integration.

There are many different companies which also have different distribution strategies. Some companies have many stages to pass through when reaching to its final customers. On the contrary, some other companies do not have many stages. Amazon.com and E-Bay are the e-business companies which eliminate the stage of distributors from their supply chains. Eliminating distributors can be cost effective but we cannot miss the certain advantages that distributors provide.

Distributors in the supply chain decrease the response time certainly. For example, Dell is a company which directly ships to the customer in an amount of time such as one or two weeks from the manufacturing stage. In such a situation, customer needs to wait and it increases the response time. However, in Radio Shack's case, the customer is able to see the products after distributors send the products to retailers. Adding more distributors may increase the product availability, lower transportation costs. The distributors increase the rate of returnability since it would be very hard to return the product to the manufacturing facility.

The distributors are able to stock inventories in reasonable amounts unless they don't increase the inventory costs dramatically. This way, they lower the risk that the retailers cannot obtain the product. It is simply called safety stock and manufacturers sometimes may not be able to keep inventories if their forecasting and information systems are not good. By having distributors, the facilities/manufacturers can focus on manufacturing instead of focusing on the distribution to retailers and/or customers. Hence, they increase their efficiency and effectiveness.

Manufacturing has switched from push system to a pull system. Forecasting demand and producing in massive amounts, replaces with producing based on the demand minimizing inventory in manufacturing plants and warehouses. Even build-to-order cycles are evolving to order-to-delivery cycles. For example, steel and paper industries get benefit from economy of scale that is mostly a feature of push systems. Economy of scale (make-to-plan strategy) suggests producing more as long as unit cost per manufactured product does not increase. High fixed cost because of manufacturing equipment exists. Economy of scale is achieved by producing specific products. In case of different products manufacturing, economy of scope employing flexible manufacturing systems is preferred. Small lot sizes of variety of products are the outputs. In fact, most manufacturing processes include combination of economy of scale and economy of scope. Capacity, equipment, and setup/changeover are the three primary constraints for manufacturing operations [2]. Manufacturing processes have impact on supply chain efficiency. Job shop process is customized for a specific need, batch process manufactures small quantities of a product before producing another one, line flow process typically uses assembly lines to build a final product merging the components of it through a line, and continuous process has a little variety such as manufacturing chemicals [2]. These processes maybe used together in a manufacturing plant. Characteristics of these processes are given in Table 1.1.

Strategy column of the table has MTO, make-to-order strategy. ATO is assembly-to-order strategy that means manufacturing components and assembling them based on customer order. MTP is make-to-plan. Total manufacturing cost includes manufacturing, inventory, and transportation costs. The relationship

Table 1.1 Characteristics of manufacturing processes [2]

Process/char.	Product variety	Volume	Strategy	Lead time
Job shop	Very high	Very low	MTO	Very long
Batch	High	Low	MTO/ATO	Long
Line	Limited	High	ATO/MTP	Short
Continuous	Very limited	Very high	MTP	Very short

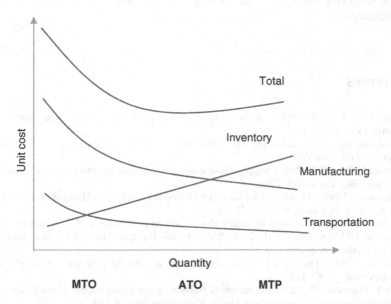

Fig. 1.2 Costs associated with manufacturing quantity [2]

between the costs regarding manufacturing quantity versus unit costs under MTO, ATO, and MTP strategies is shown in the Fig. 1.2.

It is estimated that around 90 % of global demand is not fully satisfied by local supply [2]. The cost of logistics on a global scale is estimated to exceed $8 trillion a year [2]. The authors report experts agree that logistics expenditures in the United States in 2007 were around 10.1 % of the $13.84 billion GDP. Transportation costs were 61.3 % of the total logistics cost.

Remarks

- Supply chain decisions may conflict each other that requires coordination and integration through supply chain to optimize the processes in it globally.
- Supply chain problems may require constructing mathematical models, solving the models and analyzing the results to make operational decisions.

- A supply chain can be viewed as a complex network composed of multiple echelons such as suppliers, manufacturers, warehouses, distributors, retailers, and customers.
- Today's challenge in supply chains is responding to customer need fast with a reasonable cost.
- Response time, product availability and variety, customer experience, time to market, order visibility, and returnability effect supply chain design decisions
- Manufacturing processes (i.e. batch, job shop) have impact on supply chain efficiency.

References

1. Beth S et al (2006) Supply chain challenges. Harvard business review on supply chain management, pp 65–86
2. Bowersox DJ, Closs DJ, Cooper MB (2010) Supply chain logistics management, 3rd edn. McGraw-Hill, NewYork
3. Chopra S, Meindl P (2001) Supply chain management: strategy, planning, and operations. Prentice-Hall, Englewood Cliffs
4. Ferdows K, Lewis MA, Machuca JAD (2006) Rapid-fire fulfillment. Harvard business review on supply chain management, pp 49–63
5. Hugos MH (2011) Essentials of supply chain management, 3rd edn. Wiley, London
6. Lambert DM, Knemeyer AM (2006) We're in this together. Harvard business review on supply chain management, pp 1–19
7. Lee HL (2006) The triple-a supply chain. Harvard business review on supply chain management, pp 87–115
8. Mo Y, Harrison TP (2005) A conceptual framework for robust supply chain design under demand uncertainty. Supply Chain Optim Appl Optim 98:243–263
9. Narayan VG, Raman A (2006) Aligning Incentives in supply chains. Harvard business review on supply chain management, pp 171–193
10. Spear SJ (2006) Learn in to lead at Toyota. Harvard business review on supply chain management, pp 147–169
11. Spear SJ, Bowen HK (2006) Decoding the DNA of the Toyota production system. Harvard business review on supply chain management, pp 117–145
12. Watson M, Lewis S, Cacioppi P, Jayaraman J (2012) Supply chain network design: applying optimization and analytics to the global supply chain. FT Press

Chapter 2
Supply Chain Management

Abstract Supply chain management deals with decisions on new facility locations, quantities to manufacture, modes of transporting the manufactured goods,
and information systems to use. Material and manufacturing requirements planning are conducted in a hierarchical manner. In other words, bill of materials and
master production schedule is constructed and then manufacturing orders are
released to satisfy the varying demands of the periods that are thought to be
deterministic. This chapter presents some of the important topics in supply chain
management.

Keywords Supply chain management · MRP · EOQ · Transportation

The term *supply chain management* (SCM) is attributed to Proctor and Gamble
(P&G). P&G used the term for tracking the flow of Pampers diapers through the
distribution channel [3]. As mentioned in the Chap. 1, supply chain management
deals with integration and coordination of location of facilities, production,
inventory control, and transportation of materials and products. This chapter deals
with key supply chain management decisions and planning throughout the supply
chain.

2.1 Key Supply Chain Decisions

Location of plants, warehouses, distribution centers (DCs), manufacturing quantities, order dates, inventory policies, and transportation related decisions are very
important for supply chain success. Information system employed for the supply
chain is also a key in successful implementations. These decision problems need to
be elaborated in detail.

Manufacturers face the problem of shortage in production capacity as the
demand for an item increases. The cost of outsourcing might be more than the cost
of opening a new facility or increasing the capacity of the current one by extra
labor, equipment etc. in the long run that makes opening a new facility, increasing

H. Pirim et al., *Supply Chain Management and Optimization in Manufacturing*,
SpringerBriefs in Manufacturing and Surface Engineering,
DOI: 10.1007/978-3-319-08183-0_2, © The Author(s) 2014

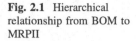

Fig. 2.1 Hierarchical relationship from BOM to MRPII

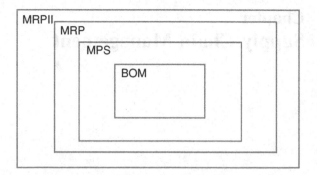

the capacity of the manufacturing plant more reasonable compared to outsourcing. Installing new machines, employment of new workers, facilitating new equipment, transportation vehicles might be necessary. Even opening a distinct plant might be compulsory. Decision on the location of the new plant based on the outbound transportation costs, operational costs within the new plant will be considered as well then. Some of the optimization problems and solutions to these problems that will lead to management decisions are reviewed in Sect. 4.2.

Location decision is a strategic one. On the other hand, manufacturers need to decide on the production quantity at operational level. Before operations level planning, aggregate planning should be achieved. Aggregate planning spans yearly plans of productions. These plans are decomposed into shorter term productions plans. Production quantity decision is complex since it comprises demand forecasts, actual demands, judgments of people from marketing, production and other departments. Capacity of the plant regarding work staff level, machine level, etc. is also a constraint for production quantity decisions. Material requirement planning (MRP) is used to decide on the production levels of end items and sub-assemblies. If the demand is known (or forecasted) and variable in each period, MRP may be employed as a top-down approach. Production planning under probabilistic stationary demand is discussed in Sect. 2.2. MRP works as a push system since it relies on forecast of the end items and production quantities push the production of sub-assemblies. MRP structure and its relation to manufacturing planning is shown in Fig. 2.1. MRP has bill of materials and master production schedule components. If capacity constraints are considered then it becomes a more global planning tool called manufacturing resource planning (MRP II) that is included in enterprise resource planning (ERP).

For example, a toy laptop consists of an assembly of a screen and lower part assembly. Lower-part assembly consists of a board on which chips are installed and a keyboard. A tree that shows the dependency between these parts is called bill of materials (BOM). BOM may be represented as a list or tree as shown in Fig. 2.2.

Lead times (LT) are given in weeks. Based on the lead times, a toy laptop is produced in 4 weeks. Table 2.1 shows the weekly demands for the next 6 weeks starting from the fifth.

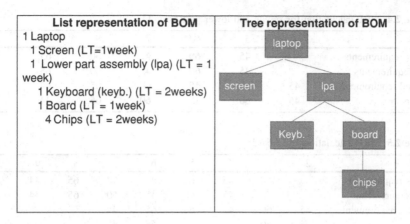

List representation of BOM	Tree representation of BOM

1 Laptop
  1 Screen (LT=1week)
  1 Lower part assembly (lpa) (LT = 1 week)
    1 Keyboard (keyb.) (LT = 2weeks)
    1 Board (LT = 1week)
      4 Chips (LT = 2weeks)	

Fig. 2.2 Bill of materials

Table 2.1 Weekly demands for toy laptop

Week 5	Week 6	Week 7	Week 8	Week 9	Week 10
50	60	45	70	78	40

Table 2.2 Scheduled toy laptop returns

Week 5	Week 8	Week 9	Week 10
5	10	13	6

Table 2.3 Master production schedule for toy laptop

Week 5	Week 6	Week 7	Week 8	Week 9	Week 10
45	60	45	60	65	44

The company might receive returns throughout 6 weeks. Let's assume scheduled receipts as given in Table 2.2.

The company updates the inventory according to scheduled receipts and it is fair to assume that at the end of the last week the company policy requires an inventory level of 10 laptops. Master production schedule is prepared netting the demand by inventory information as shown in Table 2.3.

Now these plans are pushed to next levels down the bill of materials tree. The MPS will be translated as gross requirement for lower part assembly, and screen. There is no multiplicative factor since one laptop requires one from each sub-part (screen, lower part assembly). Also, assuming that there will be no scheduled receipt and on hand inventory for the sub-parts, we can MRP calculations for both screen and lower part assembly as seen in Table 2.4.

Table 2.4 MRP calculation for screen and lower part assembly

Week	4	5	6	7	8	9	10
Gross requirements		45	60	45	60	65	44
Net requirements		45	60	45	60	65	44
Shifted requirements	45	60	45	60	65	44	
Orders	45	60	45	60	65	44	

Table 2.5 MRP calculations for board

Week	2	3	4	5	6	7	8	9	10
Gross requirements			45	60	45	60	65	44	
Net requirements			45	60	45	60	65	44	
Shifted requirements	45	60	45	60	65	44			
Orders	45	60	45	60	65	44			

Table 2.6 MRP calculations for keyboard

Week	2	3	4	5	6	7	8	9	10
Gross requirements			45	60	45	60	65	44	
Net requirements			45	60	45	60	65	44	
Shifted requirements		45	60	45	60	65	44		
Orders		45	60	45	60	65	44		

Here orders quantities are the same with the lead time shifted requirements. That is known as lot-for-lot ordering policy. Amount of order may differ based on different ordering policies. Some of them are reviewed in Sect. 2.2.

Similar calculations are made for board and keyboard. MRP calculations are shown in Tables 2.5 and 2.6.

Assuming that chips are similar to each other, one board requires four chips. So ordering boards starting from third week pushes chip orders 2 weeks before with the quantity of four times the amount of boards. Table 2.7 shows the MRP calculations.

Here, demands are assumed to be deterministic. In reality, manufacturers resort to safety stocks because of the uncertainty in demands. If we approximate the cumulative distribution value for meeting the demand, i.e. normally distributed, we can add safety stock to our demands to be used as new gross requirements. For example, if we want to meet the demand (normally distributed) for toy laptop each week with a probability of 95 %. Then average demand + standard deviation times 1.65 (standard normal variate value) will give the new gross requirements.

Lead times might not be deterministic as well. They also can be adjusted, for example by a multiplicative factor to include variability.

Capacity of the plant may be a constraint to produce the orders from MRP. Capacity planning shifts MRP to MRP II (manufacturing resource planning)

Table 2.7 MRP calculations for chip

Week	1	2	3	4	5	6	7	8	9	10
Gross requirements			180	240	180	240	260	176		
Net requirements			180	240	180	240	260	176		
Shifted requirements	180	240	180	240	260	176				
Orders		180	240	180	240	260	176			

paradigm that incorporates different departments of the company for production planning. Capacity planning problem will be reviewed in Sect. 2.3.

MRP serves as a tool to make production quantity decision. However, MRP assumes deterministic demands subject to changes in different periods. MRP is a push system. The example above assumes a static MRP that has a fixed planning horizon, 6 weeks. In reality an MRP needs to be run each period to manipulate productions decisions. *Rolling horizon* approach implements only the first-period decision of *N*-period problem [3]. When using rolling horizon approach, number of periods should be long enough to make the first-period decision constant.

2.2 Ordering Policies

In this chapter, MRP calculations resulted in number of orders and we determined the number based on a lot-for-lot policy. Order lot size is equal to the lead time shifted requirements. However, this lot sizing policy is not necessarily optimal. There are other order size policies and also there is an optimal policy.

The simplest model to start is for the uncapacitated single item lot sizing problem (USILSP). A natural mixed integer formulation of the problem is given as follows [1]:

$$\min \sum_{t=1}^{T} (s_t Y_t + c_t X_t + h_t I_t)$$

subject to

$$I_{t-1} + X_t - D_t = I_t; \forall t$$

$$X_t \leq Y_t D_{tT}; \forall t$$

$$Y_t \in \{0, 1\}; \forall t$$

$$X_t, I_t \geq 0; \forall t$$

s_t is the set-up cost in period t ($t = 1,\ldots,T$). c_t is unit production cost in period t. h_t is inventory holding cost in period t. X_t is the production quantity in period t. I_t is the inventory at the end of period t. $D_{tT} = D_t + D_{t+1} + \cdots + D_T$. Here, beginning and ending inventory levels are zero.

The objective function of the model minimizes the total cost that includes set-up cost at each production run, production cost, and inventory cost over T periods. First set of constraints imply that the inventory level at the end of period t is equal to the sum of inventory level of the previous period and production amount in period t minus the demand in the same period.

The model can be extended to include multiple facilities introducing W_{jkt} transfer variables defined as quantity transferred from facility j to facility k in period t. The new objective function includes transfer cost and inventory constraints include transferred products:

$$\min \sum_{j}^{F} \sum_{t=1}^{T} \left(s_{jt} Y_{jt} + c_{jt} X_{jt} + h_{jt} I_{jt} + \left(\sum_{k \neq j} r_{jkt} W_{jkt} \right) \right)$$

subject to

$$I_{jt-1} + X_{jt} + \sum_{l \neq j} W_{ljt} - D_{jt} = I_{jt} + \sum_{k \neq j} W_{jkt}; \forall j, t$$

$$X_{jt} \leq Y_{jt} \sum_{j=1}^{F} \sum_{i=t}^{T} D_{ji}; \forall j, t$$

$$Y_{jt} \in \{0, 1\}; \forall j, t$$

$$X_{jt}, I_{jt}, W_{jkt} \geq 0; \forall j, \quad k \neq j, t$$

Capacity constraints can be added to both of the models introduced above.

Since integer programming models are hard to solve, it might be efficient to use heuristics to find a reasonable—not optimal solution to a lot sizing problem. Here are some of the widely used ones:

1. Silver-Meal heuristic
2. Least unit cost heuristic
3. Part period heuristic

Silver-Meal is a myopic heuristic that works based on average cost per period. The cost function of the heuristic spans future periods as long as the value of it increases. $C(t, t + n)$ is the cost in period t to cover periods from t to $t + n, n + 1$ periods. D_t is the demand in period t, then the cost spanning $n + 1$ periods is found by:

$$C(t, t+n) = S + h \sum_{i=0}^{n} i D_{t+i}$$

The first period cost $C(1, 1)$ is only the set-up (or order) cost S. The average cost spanning two periods is:

$$\frac{C(1,2)}{2} = \frac{S + h\sum_{i=0}^{1} iD_{t+i}}{2} = \frac{S + hD_2}{2}.$$

The average cost spanning three periods is:

$$\frac{C(1,3)}{3} = \frac{S + hD_2 + 2hD_3}{3}.$$

As we generalize it:

$$\frac{C(1, n+1)}{n+1} = \frac{S + hD_2 + 2hD_3 + \cdots + nhD_{n+1}}{n+1}.$$

The stopping criteria for the heuristic is

$$\frac{C(t, t+n)}{n+1} > \frac{C(t, t+n-1)}{n}.$$

Once the heuristic stops, the lot size for period t is set as $D_t + D_{t+1} + \cdots + D_{t+n-1}$ and the heuristic starts over at period $n + 1$.

If we return to our toy laptop example in this chapter, shifted requirements for laptop screen were 45, 60, 45, 60, 65 and 44. Let's assume an $400 order cost for screens and holding cost of $5. Then we can work out Silver-Meal heuristic.

$$C(1,1) = 400, \frac{C(1,2)}{2} = \frac{400 + 5 \times 60}{2} = 350,$$

$$\frac{C(1,3)}{3} = \frac{400 + 5 \times 60 + 2 \times 5 \times 45}{3} = 383.33$$

We set the lot size for period one as $45 + 60 = 105$ and start over from third period.

$$C(3,3) = 400, \frac{C(3,4)}{2} = \frac{400 + 5 \times 60}{2} = 350,$$

$$\frac{C(3,5)}{3} = \frac{400 + 5 \times 60 + 2 \times 5 \times 65}{3} = 450$$

We set the lot size for period three as $45 + 60 = 105$ and start over from fifth period.

$$C(5,5) = 400, \frac{C(5,6)}{2} = \frac{400 + 5 \times 44}{2} = 310$$

Since all periods are over we set the lot size for period five as $65 + 44 = 109$.

We can make cost comparison between lot-for-lot policy and Silver-Meal policy. Lot-for-lot policy will have only order costs of $6 \times 400 = \$2,400$. Silver-Meal will have order costs of $3 \times 400 = \$1,200$ and holding costs of $300 + 300 + 220 = \$820$. Total cost is $2,020. So, Silver-Meal saves around 16 % here.

Least unit cost heuristic can be viewed as a modified version of Silver-Meal heuristic. Modification is made on the cost function. The cost function is divided by the total demand, instead of number of periods.

We can write unit cost expressions for the first period spanning one period as:

$$\frac{C(1,1)}{D_1} = \frac{S}{D_1}.$$

The unit cost expression spanning two periods starting from the first one is

$$\frac{C(1,2)}{D_1 + D_2} = \frac{S + hD_2}{D_1 + D_2}.$$

The unit cost expression spanning three periods starting from the first one is

$$\frac{C(1,3)}{D_1 + D_2 + D_3} = \frac{S + hD_2 + 2hD_3}{D_1 + D_2 + D_3}.$$

General unit cost expression spanning $n + 1$ periods starting from the first one is

$$\frac{C(1, n + 1)}{D_1 + \cdots + D_{n+1}} = \frac{S + hD_2 + 2hD_3 + \cdots + nhD_{n+1}}{D_1 + \cdots + D_{n+1}}.$$

Stopping criteria for the heuristic is:

$$\frac{C(t, t + n)}{D_t + \cdots + D_{n+1}} > \frac{C(t, t + n - 1)}{D_t + \cdots + D_n}.$$

The lot size for period t is set as $D_t + D_{t+1} + \cdots + D_{t+n-1}$ and the heuristic starts over at period $n + 1$.

We can apply the unit cost heuristic to the same example:

$$\frac{C(1,1)}{D_1} = \frac{400}{45} = 8.88, \frac{C(1,2)}{D_1 + D_2} = \frac{400 + 5 \times 60}{105} = 6.66,$$

$$\frac{C(1,3)}{D_1 + D_2 + D_3} = \frac{400 + 5 \times 60 + 2 \times 5 \times 45}{150} = 7.66.$$

Stopping criteria is met. Lot size for the first period to span two periods is $45 + 60 = 105$. We start over from the third period:

$$\frac{C(3,3)}{D_3} = \frac{400}{45} = 8.88, \frac{C(3,4)}{D_3 + D_4} = \frac{400 + 5 \times 60}{105} = 6.66,$$

$$\frac{C(3,5)}{D_3 + D_4 + D_5} = \frac{400 + 5 \times 60 + 2 \times 5 \times 65}{170} = 7.94.$$

The lot size for the third period to span two periods is $45 + 60 = 105$. Starting over from fifth period:

$$\frac{C(5,5)}{D_5} = \frac{400}{65} = 6.15, \frac{C(5,6)}{D_5 + D_6} = \frac{400 + 5 \times 44}{109} = 5.69.$$

The unit cost heuristic stops since the number of periods is reached. The lot size for the fifth period to span two periods is $65 + 44 = 109$. The lot sizes are the same with Silver-Meal results. However, it is most likely that two heuristics will result in different lot sizes solving bigger real world problems. Heuristics are not guaranteed to find optimal solutions. Also, it is hard to judge which heuristic is better for all scenarios.

Part period heuristic aims to balance set-up cost and inventory holding cost. Assuming the inventory holding cost $I(t, t + n)$ associated with carrying inventory for n periods. If the inventory holding cost is greater than the set-up cost, then it is reasonable to place a new order at the period $t + n$.

Using the data for the toy laptop example, the first period will not have any inventory holding cost, $I(1, 1) = 0$. The holding cost for carrying from first to second period $I(1, 2)$ will be $5 \times 60 = 300$ that is less than the set-up cost. The holding cost carrying till third period $I(1, 3)$ will be $5 \times 60 + 2 \times 5 \times 45 = 750$ that is more than the set-up cost, 400. So, we set the lot size for the first period $45 + 60 = 105$, and place a new order for the third period. Holding cost for the third period $I(3, 3)$ will be zero. $I(3, 4) = 5 \times 60 = 300$ that is less than the set-up cost. $I(3, 5) = 5 \times 60 + 2 \times 5 \times 65 = 950$ that is more than the set-up cost. The lot size for the third period to cover two periods is $45 + 60 = 105$. We place a new order for the fifth period and the lot size is calculates as follows: $I(5, 5) = 0$, $I(5, 6) = 5 \times 44 = 220$ that is less than the set-up cost. The heuristic stops since the number of periods is reached. The lot size for the fifth period is $65 + 44 = 109$.

For this problem three of the heuristics gave the same result associated with a total cost value of \$2,020.

Besides IP models and heuristic methods, dynamic programming approaches are used for lot sizing as well. Dynamic programming breaks the problem into overlapping sub-problems, solves each sub-problem optimally and uses these solutions for finding the optimal solution to the original problem. Here, finding the optimal lot sizes can be represented as a directed acyclic network. Then, the shortest path on the acyclic network gives the optimal solution, lot sizing policy. Dynamic programming can be employed to find the shortest path on the directed acyclic network. Nodes of the network represent the periods. An extra node is added to represent the end of periods. Arc (i, j) represents that ordering happens at period i and the lot size is $D_i + D_{i+1} + \cdots + D_{j-1}$ and next ordering happens at period j. The network for the toy laptop example is shown in Fig. 2.3.

For example, if the optimal lot sizing policy required ordering in the first, third, and the fifth period that would mean path 1–3–5–7 (for toy laptop example, we need seven nodes). Arc weights (c_{ij}) are the costs that include set-up and/or inventory holding cost. C_{ij} is defined as the cost of ordering in period i to cover

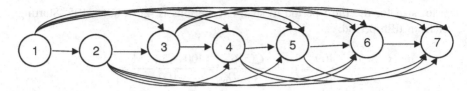

Fig. 2.3 Directed acyclic network for the toy laptop example

demand through period $j - 1$. Let f_i be the minimum cost starting at node i with the order placed in period i. Then we define a recursion:

$$f_i = \min(c_{ij} + f_j), \quad i < j, \quad i = 1, \ldots, n$$

The minimum cost for the ending node is zero, $f_{n+1} = 0$.

Our example has six period, seven nodes, $f_7 = 0$.

$$f_6 = \min(c_{6j} + f_j) = 400. \text{ Here } j \text{ can only take the value seven.}$$

$$f_5 = \min(c_{5j} + f_j) = \min\left\{ \begin{array}{c} c_{56} + f_6 \\ c_{57} + f_7 \end{array} \right\} = \min\left\{ \begin{array}{c} 400 + 400 \\ 620 + 0 \end{array} \right\} = 620, \quad j = 7$$

$$f_4 = \min(c_{4j} + f_j) = \min\left\{ \begin{array}{c} c_{45} + f_5 \\ c_{46} + f_6 \\ c_{47} + f_7 \end{array} \right\} = \min\left\{ \begin{array}{c} 400 + 620 \\ 725 + 400 \\ 1{,}165 + 0 \end{array} \right\} = 1{,}020, \quad j = 5$$

$$f_3 = \min(c_{3j} + f_j) = \min\left\{ \begin{array}{c} c_{34} + f_4 \\ c_{35} + f_5 \\ c_{36} + f_6 \\ c_{37} + f_7 \end{array} \right\} = \min\left\{ \begin{array}{c} 400 + 1{,}020 \\ 700 + 620 \\ 1{,}350 + 400 \\ 2{,}010 + 0 \end{array} \right\} = 1{,}320, \quad j = 5$$

$$f_2 = \min(c_{2j} + f_j) = \min\left\{ \begin{array}{c} c_{23} + f_3 \\ c_{24} + f_4 \\ c_{25} + f_5 \\ c_{26} + f_6 \\ c_{27} + f_7 \end{array} \right\} = \min\left\{ \begin{array}{c} 400 + 1{,}320 \\ 625 + 1{,}020 \\ 1{,}090 + 620 \\ 2{,}065 + 400 \\ 2{,}945 + 0 \end{array} \right\} = 1{,}635, \quad j = 4$$

$$f_1 = \min(c_{1j} + f_j) = \min\left\{ \begin{array}{c} c_{12} + f_2 \\ c_{13} + f_3 \\ c_{14} + f_4 \\ c_{15} + f_5 \\ c_{16} + f_6 \\ c_{17} + f_7 \end{array} \right\} = \min\left\{ \begin{array}{c} 400 + 1{,}635 \\ 700 + 1{,}320 \\ 1{,}150 + 1{,}020 \\ 2{,}050 + 620 \\ 3{,}350 + 400 \\ 4{,}450 + 0 \end{array} \right\} = 2{,}020,$$

$$j = 3$$

To obtain lot sizes, we backtrack the solution. Last solution informs that $j = 2$, lot size is equal to the first period's demand, 45. Next order is in period two, where j value is four. So, the lot size will cover demands for periods two and three that is 105. Next order is in period four, where j value is five. The lot size is equal to the demand in period four, 60. The next order is in period five, where j value is seven. The lot size will cover demands for periods five and six that is 109.

So the optimal solution is the path 1–3–5–7. Lot sizing policy is ordering 105 in the first period, 105 in the third period and 109 in the fifth period. As seen in the results before, heuristics also found the optimal solution for this example.

Till here, we assumed deterministic demands. However in real world scenarios, it is highly likely that demand changes fitting a statistical distribution. Newsboy model is a widely used approach. We can assume the demand D as a random variable. A boy purchases Q newspapers to sell and based on the demand, he has an underage cost c_u (when demand is more than the number of newspapers, Q) or overage cost c_o (when Q is greater than the demand). Then the optimal number of newspapers to purchase is found by:

$$F(Q) = \frac{c_u}{c_u + c_o}$$

Here, $F(Q)$ is the cumulative distribution function of demand at Q. That's the probability that the demand is less than Q.

Lot size re-order systems reviews the system continuously. The system has two variables R and Q. When inventory level hits R, Q units are ordered. As we assume a lead time L, demand during the lead time becomes the source of uncertainty. S is the set-up cost, p is the penalty cost per unit for unsatisfied demand. Then the following equations are solved back and forth iteratively [3]:

$$Q = \sqrt{\frac{2D[S + pn(R)]}{h}},$$

$$1 - F(R) = \frac{Qh}{ph}.$$

F(R) is the cumulative distribution function of D. One approximation is setting Q value to EOQ value and solving it for R. n(R) is the expected number of shortages in a cycle:

$$E(\max(D - R, 0)) = \int_{R}^{\infty} (x - R)f(x)dx$$

(Q, R) values are found through continuous review policy. In periodic review systems (s, S) policy is used. When the inventory on hand is less than or equal to s, quantity up to S is ordered.

2.3 Capacity Planning

Demands may not be able to be satisfied each period because of some capacity restrictions. Even, lot size decisions may not be feasible because of the capacity constraints. Considering toy laptop example, D (here, net requirements) = (45, 60, 45, 60, 65, 44) we can assume that production capacities for each period Cap = (50, 50, 50, 50, 50, 50). The following constraints must be satisfied to maintain feasibility.

$$\sum_{i=1}^{j} Cap_i \geq \sum_{i=1}^{j} D_i; \quad j = 1, \ldots, 6$$

We can check if the problem is feasible.

First period constraint: $50 \geq 45$ is satisfied.

Second period constraint: $100 \leq 105$ is not satisfied. We don't need to check remaining constraints since the problem became infeasible. We cannot satisfy the demands of the first two periods with our available resources for the first two periods. However, all of the constraints were satisfied, then the next step would be to find an initial feasible solution. For example, as we increase the capacities for each period to 60, the problem becomes feasible. We can shift back demands to find initial solution. Fifth period net requirements is more than our capacity, so five units are shifted to third period. Then our new production/ordering schedule becomes $D' = (45, 60, 50, 60, 60, 44)$. Now we can improve the initial solution. There may be different approaches to improve the solution, we adopt one mentioned by Nahmias [3]. The idea is to shift production orders back as long as the holding costs is less than the set-up costs starting from the last period. In our example, we don't have enough capacity in previous periods to shift 44 back.

Production decisions may change based on the structure of the demand (deterministic vs. stochastic, stationary). Inventory review policies (periodic review vs. continuous review) may affect the production decisions as well.

Inventory policy decisions is based on the costs associated with holding inventory and set-up costs. Economic order quantity (EOQ) model is a simple approximation for a quantity decision based on total production cost. The simplest EOQ model assumes that demand rate is constant. Once the order of Q is given (when the inventory level hits zero), the inventory level is updated to Q immediately. In other words, the model assumes lead time zero. Shortage is not allowed. Each order has a fixed set-up cost of S, variable cost of c per unit, and a holding cost h per unit per inventory holding time is charged. Usually holding cost is expressed as a percentage of c. The objective is finding the Q level that will minimize the average production cost per period (usually a year). Each ordering cycle will have a cost of $S + cQ$. Assuming that cycle length is L, dividing the cost expression by L will give the cost per unit time. Q units are used by demand rate D. Hence, $L = Q/D$. The average inventory level per cycle is $Q/2$ since Q decreases linearly. Then, we compute average annual (periodical in general) cost (AAC) as:

$$\text{AAC(Q)} = \frac{S+cQ}{L} + \frac{hQ}{2} = \frac{S+cQ}{Q/D} + \frac{hQ}{2} = \frac{SD}{Q} + Dc + \frac{hQ}{2}$$

Last three terms include average periodical set-up cost, purchase cost, and inventory cost. The cost function is convex function. Hence, the Q value based on the first derivative of the expression will be the global optimum. In other words, Q value that satisfies $\text{AAC(Q)}' = 0$ is the optimal value denoted as Q^* known as EOQ. The EOQ formulation is:

$$Q^* = \sqrt{\frac{2SD}{h}}$$

For example, if the weekly demand for laptop toy is 500 units and set-up cost to initiate the order is \$200, and a laptop has a variable cost of \$5 per unit, assuming a holding cost of 10 % of variable cost per period, we can find the optimal order quantity:

$$Q^* = \sqrt{\frac{2 * 200 * 500}{0.5}} \cong 633$$

Here, set-up cost is relatively high compared to holding cost. It is reasonable to order in high quantities once every 9 days (663/500 translated to days, assuming 7 days a week). Since set-up cost are usually high in batch or mass production, this example also shows that to achieve just in time (JIT) production or eliminate inventory set-up time reductions (assuming set-up costs are proportional to set-up time) is a critical point. As JIT requires frequent orders of small batch sizes.

After decision of order or production quantities, transportation decisions should be made. The company may have a contract with third party carriers or may use its own trucks and transportation facilities to deliver products to customers. Especially, international firms need to consider modes of transportation, inbound and outbound logistics costs. Road, railway, waterway, air, and pipelines are common modes of transportation. Intermodal transportations are possible as well.

Road transportation is preferred inside a country. The main rule is to be able to carry as long and as much as possible to minimize the transportation cost. Monitoring this mode of transportation is easy. Perishable and non-perishable items may be carried. Some disadvantages are: there may be delays due to traffic, some regulations may be a restriction on driving routes, might be affected by weather conditions and subject to accidents that will lead to severe damages on products.

Railway transportation has a capacity and cost advantage compared to road transportation. Even, it is safer and more reliable. A disadvantage is that railways are limited worldwide and rail freight destinations may be far away from customer. Hence, delivery to customer needs to be handled after railway transportation.

Waterway is used to carry heavy and huge items. This mode of transportation is slow and may be cheap compared to road and railway. Disadvantages are long lead times, subject to bad weather influence, inter-country restrictions are available.

Table 2.8 Comparison of transportation modes

Mode	Intercity tonnage	Intercity ton-miles	Freight expense	Revenue
Road	3,745	1,051	402	9.1TL, 26.1LTL
Railway	1,972	1,421	35	2.4
Waterway	1,005	473	25	0.7
Air	16	14	23	56.3
Pipeline	1,142	628 (oil)	9	1.4

Air transportation is the fastest and the most expensive mode of transportation. Air transportation is due to flight schedule cancellations or changes and may have restriction on items to deliver.

Pipeline transportation is used for transferring gas, petroleum products, and sewage. The flow is slow, and investment cost is high. However, this mode is not affected by weather conditions and flow goes on continuously. Pneumatic tubes are used for example in hospitals to deliver documents, blood samples etc.

Chopra [2] gives the intercity weight (in millions of tons) and distance (in billions of ton-miles) capacities, freight expenses (in billions of dollars) and revenue (cents per ton-mile) in US shown in Table 2.8.

Of course transportation costs may affect facility location decisions. Review of some optimization problems regarding transportation is in Sect. 4.1.

Remarks

- Key supply chain management decisions include selection of new facility locations, manufacturing quantities, transportation, and information system related decisions.
- MRP is a push system that deals which resource planning in a hierarchical manner. Running MRP system relies of bill of materials and master production schedule. Demands are viewed as deterministic, varying by period.
- Different lot sizing policies exist. Integer programming formulation for the uncapacitated single item lot sizing problem gives the optimal solution.
- Heuristic approaches include Silver-Meal, unit cost, and part period heuristics.
- Lot sizing can be represented as a directed acyclic network. Dynamic programming may be employed to find the shortest path of the network that is the optimal lot sizing policy.
- Newsboy model is used in periodic review problems. It ignores set-up cost.
- (Q, R) policy requires continuous review. Once the inventory level hits R, Q quantity is ordered. In periodic review (s, S) policy $S–I$ is ordered if the inventory on hand (I) is less than or equal to s.
- Economic order quantity (EOQ) model assumes a constant demand rate. Shortage in fulfilling orders is not allowed.
- Different modes of transportation have benefits and disadvantages and they have an effect on supply chain success.

References

1. Brahimi N et al (2006) Single item lot sizing problems. Eur J Oper Res 168:1–16
2. Chopra S, Meindl P (2004) Supply chain management, 2nd edn. Pearson, Upper Saddle River
3. Nahmias S (2009) Production and operations analysis, 6th edn. McGraw Hill, New York

Chapter 3
Scheduling Models in Supply Chain

Abstract This chapter introduces the scheduling models in supply chains. Models of scheduling within production units are discussed for different shop structures and objectives. Such models and solution methods are used as a base for further development across production units with objectives of increasing the synergy resulting from coordinated or integrated scheduling. The chapter briefly introduces some of the basic models in scheduling theory that mostly related to supply chain models followed by some of the basic models in supply chain scheduling.

Keywords Supply chain · Scheduling · Mathematical models · Optimization

3.1 Introduction

Production scheduling is an important element in optimizing resource utilization and satisfying customer timely needs. Optimum scheduling under different conditions, deterministic and stochastic, for different objectives internal and external, are well studied in the literature for different settings, discrete and continuous, and simple and complex production structures. Single machine, flaw shop, job shop, and flexible production systems are considered for optimally scheduling jobs. Dynamic and static scheduling for different types of production systems is also considered. These studies were conducted in the last five decades.

Advances in communication and information technology and global international relations have changed the way business is conducted around the globe. Business giants are emerging, the economy of scale is prevailing, and global competing is getting tough as it has never been. In response to these changes, supply chains are getting more global raising the need for research in global supply chain business environment. Transportation, inventory, production and delivery scheduling are some of the areas that need to be studied for optimized sharing benefits among the supply chain partners from raw material suppliers to the end users. Production scheduling is a major decision that affects all resources including inventory, transportation as well as human resources. The scheduler in each

H. Pirim et al., *Supply Chain Management and Optimization in Manufacturing*,
SpringerBriefs in Manufacturing and Surface Engineering,
DOI: 10.1007/978-3-319-08183-0_3, © The Author(s) 2014

production unit in the supply chain should take into consideration the production schedule of the upstream customers and the downstream supplier chains in addition to his own internal production conditions and limitations. Reduction in production in a production unit supplying parts to another unit due to maintenance needs will disturb the production in that unit and the effect will propagate until it reached the end customer. In this case all producers in the supply chain will be affected. This negative effect could have been avoided if that reduction was planned in coordination with upstream units in which case other sources could have been arranged or production schedule adjusted to accommodate for the anticipated shortage.

At the operational level, decision makers at different stages of the chain need to consider their immediate customers' due dates, and production deadlines, changeover costs and times. As a result, each stage defines its own ideal schedule that specifies how orders should be processed at that stage. For example, an assembly facility which has to ship jobs to different customers may wish to process the materials in the same sequence as the due dates. On the other hand, according to JIT concepts, scheduling decisions at an upstream stage must also comply with the actual time at which the supplier will dispatch the raw materials and with technological requirements that may make certain schedules infeasible. Thus, the schedule that is used at each stage depends on the requirements at the other stages [1].

Such conflicting decisions at different levels in the supply chain raise the issue of coordination in supply chain decisions, including production planning and scheduling. Hall and Potts [4] demonstrated through three examples that the solution which results from the supplier and manufacturer acting independently is considerably more costly than the solution of the combined problem. The examples showed that cooperation between supplier manufacturers may reduce the total system cost by 20–25 % and may go up to 100 %, depending on the scheduling objective.

This chapter considers production scheduling optimization in a supply chain context with multiple production units. Production schedule at each production unit in the supply chain need to consider its customer's due dates, production and stocking capacities, and production flexibility. The foundation of supply chain scheduling is the classical machine scheduling theory, which studies scheduling decisions in single production units of different production structures. Thus, classical scheduling models will be discussed through few examples in the next section. The following section introduces some scheduling models of different structures of supply chains.

3.2 Scheduling in Production Units

There are several types of modern manufacturing systems including intermittent, continuous and flexible production systems. Intermittent production is where more than one of the same product is being made in a short amount of time. There are structures of intermittent systems including batch production, jobbing production, and project production. In Batch production a group of similar products (batch) are produced stage by stage over a series of workstations. Batch production has a relative low initial set up cost for single production line used to produce several products. Jobbing production is where firms produce items that meet the definite requirements of the client. These items are designed differently, and are tailored to the needs of each individual client.

In project production a complex sets of interrelated activities (project) are performed within a given period of time and estimated budget to make a product characterized by its immobility during production. Examples of such products are; ships, locomotive, aircrafts buildings and bridges. The product is located in a fixed position where production resources are moved to it.

The most flexible and responsive to changes manufacturing system is the flexible manufacturing system (FMS). It absorbs sudden large scale changes in production volume, capacity and capability. FMS produces a product just like intermittent manufacturing and is continuous like continuous manufacturing. Flexibility is coming from either the ability to produce new products (machine flexibility) or from the ability to use multiple machines to perform the same operation (routing flexibility).

Continuous manufacturing is the type of manufacturing system that uses an assembly line or a continuous process to manufacture products. It is used for products that are made in a similar manner. In this type of manufacturing system the product moves and processed along the production line. Continuous processing is a method used to manufacture or process materials that are either dry bulk or fluid continuously through a certain chemical reaction or mechanical or heat treatment. Continuous usually means several months or sometimes weeks without interruption. Some common continuous processes are: Oil refining, Chemical and petrochemicals plants, sugar mills, blast furnace, power stations, and saline water desalination and cement plants. Continuous processes use process control to automate and control operational variables such as flow rates, tank levels, pressures, temperatures and machine speeds.

Planning for production goes through several levels of decisions at different time spans. Figure 3.1 represents a generic manufacturing environment flow of information and decision levels. Production planning master scheduling is a long term planning that decides on the production level for the next year in terms of type and amount of production of each type. It is based on a forecasted demand based on actual orders or estimated orders of business partners. Orders are usually associated with dates of delivery and estimated orders are also associated with estimated due dates. Based on the forecasted demand material requirements and production capacity are identified and planned for. Materials include raw material and assembly parts that need to be available at the time designated for production. Capacity is determined by human resources, machines, equipment, working hours, etc. Having the needed capacity for production available and the needed materials arranged, orders are ready for production overtime. At the scheduling stage, the forecasted orders into jobs and setting the time for producing these jobs is the scheduling stage. Scheduling jobs on machines for achieving a certain objectives

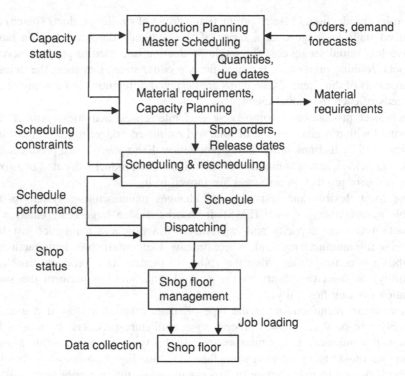

Fig. 3.1 Information flow diagram in a manufacturing system [9]

related to customer requirement as well as production, inventory, and delivery costs. Scheduling is highly linked to business upstream and downstream partners in relation to material delivery from the suppliers and product delivery to customers. This link makes the scheduling stage critical to healthy supply chain relationship. This task directly affects and affected by business partners. At the designated time jobs are dispatched to the shop floor for production. Once the job is dispatched it becomes difficulty to change, cancel or modify.

The timing of production and product delivery is determined by the two stages, master scheduling and scheduling and rescheduling. The interaction within the supply chain in relation to requirement and delivery timing goes through these two stages in each unit in the supply chain. Scheduling techniques within the production unit has long history of advancements and large volume of research and development. Scheduling within supply chain is relatively new research area with limited application. However, the accumulated knowledge in production scheduling can be used as a base for development in the area of supply chain scheduling.

3.2.1 Machine Scheduling Modelling

Several machine structures are considered in the theory of scheduling including the following:

- Single machine
- Parallel machine
- Flowshop and Jobshops
- Flexible flowshop and jobshop

The single machine scheduling is the most basic structure for which varieties of models are developed for different production conditions and constraints. Parallel machine scheduling is similar to the single machine in that it is a single stage processing environment except that it includes multiple machines in parallel. Flow shops and job shops are multiple stage production environments with subsequent operations of the same ordered requirements (flow shop) or various ordered requirements (job shop). Flexible flow shops or job shops are different by having multiple machines in parallel at each stage of production. Each of these shop structures are studied and modeled in the literature and used in practice for different conditions, constraints, and assumptions. The assumptions are in many cases related to availability of job related information such as processing times, arrival time, due date, etc. Other assumptions may be related to machines in terms of their availability, speed, quality, and capability. Constraints can be related to job batching, job interrelation that need to be considered in scheduling jobs on machines.

Various performance measures objective functions are considered in modeling and optimizing scheduling problems under the above structures. Measures might be related to machine utilization such as the makespan (the time needed to complete processing all jobs) or related to customer due date requirements such as average or maximum tardiness and number of jobs missing their due dates. Performance measures can also be related to inventory size such as average earliness and average completion time. Scheduling models for optimizing schedules with respect to each of these performance measures or with respect to more than one are developed in the literature and utilized in practice in various industries.

Several optimization techniques are used in production scheduling. Integer programming, dynamic programming, and branch and bound are commonly used in scheduling theory. Stochastic programming and simulation is mainly used under stochastic conditions related to job arrival, machine availability, or processing times. Heuristics are used for large scale problems and for handy and simple solutions. Met heuristics such as, simulated annealing, Tabu search, genetic algorithms, Ant colony, etc. are developed for various scheduling problems.

The flowshop problem is the simplest structure that resembles a simple supply chain structure. The problem is extensively studied in the literature with various models of different types and efficiencies developed and examined for different objective functions and constraints. Integer programming was one of the first models developed for optimizing flow shops.

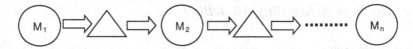

Fig. 3.2 A basic flow shop production scheme

3.2.2 Flow Shop Scheduling Problem

Flow shop production structure is composed of n stages of processing units, M_1, M_2, ..., M_n, in which different jobs pass through in the same sequence (Fig. 3.2). A job j in each stage i is processed for a certain time p_{ij}. The requirement is to determine the order of processing the n jobs starting at time zero with no interruption in a minimum time span (makespan).

In this specific problem it is assumed that machines are always available for processing and jobs are ready at the beginning of the planning horizon. It is also assumed that a buffer of unlimited capacity is available between stages to absorb accumulated jobs waiting for a machine to be released. Further assumptions include the following:

- Machines can process a single job at a time
- A job is an entity that cannot be split into sub jobs
- Transportation time between machines is negligible
- All data including number of jobs, numbers of machines, processing times are known in advance
- Once a machine start processing a job it must finish it without interruption or cancellation

The optimum schedule is composed of a sequence of jobs with the job in the first position processed starting at time zero followed by the next jobs as soon as the required machine is free. Thus each sequence produces a corresponding schedule. Gantt charts are used in representing job processing overtime for machines. An example of a Gantt chart is shown in Fig. 3.3 for a three machine four job flow shop problem for a given sequence. Each job in the chart is given a different color and width of each block represents the processing time of each operation. Changing the coloring order (job order) will change the makespan shown in heavy line in the figure. Finding the optimum order among all possible orders (sequences) requires a sophisticated analytical tool.

This problem is one of the basic flow shop production structures and one of the early mathematically modeled scheduling problems. A Mixed Integer Programming (MIP) model was developed by Wagner [11]. The model (adopted from French [3]) is as follows.

Define the following variables;

Fig. 3.3 Gantt chart for a flow shop sequence

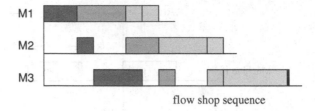

flow shop sequence

$$x_{jk} = \begin{cases} 1 & \textit{if job j is scheduled in the kth position of the proces} \sin g \textit{ sequence} \\ 0 & \textit{otherwise} \end{cases}$$

where x_{jk}, are non-negative integers taking the values 0 and 1 only.

To insure that exactly one job is scheduled in position k the following must be satisfied.

$$\sum_{j=1}^{n} x_{jk} = 1 \quad k = 1,\ldots,n$$

And to insure that each job is scheduled in exactly one position, the following must be satisfied.

$$\sum_{k=1}^{n} x_{jk} = 1 \quad j = 1,\ldots,n$$

Also define I_{ik}, $i = 1,\ldots,m$, $k = 1,\ldots,n-1$, variables representing the idle time of machine i after the completion of the kth job in the sequence. The idle times before the first job in all machines are assumed to be zero, i.e., $I_{1k} = 0$, $k = 1,\ldots,n-1$.

And define W_{ik}, $i = 1,\ldots,m-1$, $k = 1,\ldots,n$, variables representing the waiting time of the job in position k after its completion on machine i and before starting the next machine. The waiting of the first job in all machines are assumed to be zero, i.e., $W_{i1} = 0$, $i = 1,\ldots,m-1$.

Also define Δ_{ik}, as the times between the completion of the job in position k on machine I and the start of the next job $k + 1$ on machine $i + 1$ must be well defined. This is shown in Fig. 3.4.

The time Δ_{ik} should satisfy the following equality in order to avoid job overlaps (processing two jobs at the same time on the same machine).

$$\Delta_{ik} = I_{ik} + p_{i(k+1)} + W_{i,k+1} = W_{ik} + p_{i+1(k)} + I_{i+1,k}$$

Since processing times are expressed in terms of job's position in the sequence it is needed to express them in terms of the jobs absolute numbering. Processing times can be redefined as follows:

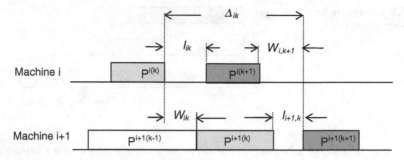

Fig. 3.4 The relationship between the variables in the model

$$P_{i,(k)} = \sum_{j=1}^{n} x_{jk}p_{ij}$$

The objective function, minimizing the makespan, is equivalent to minimizing the total time on the last machine which is given by the sum of the inter job idle times I_{mk} plus the idle time that must occur before the first job on that machine. Thus it is required to minimize

$$Makespan = \sum_{i=1}^{m-1} P_{(1)i} + \sum_{j=1}^{n-1} I_{mk}$$

Rewriting the makespan in terms of job absolute numbering yields the following,

$$Makespan = \sum_{i=1}^{m-1} \sum_{j=1}^{n} x_{j1}p_{ij} + \sum_{k=1}^{n-1} I_{mk}$$

So, the flow shop problem is formulated as follows:

Minimize $\displaystyle\sum_{i=1}^{m-1} \sum_{j=1}^{n} x_{j1}p_{ij} + \sum_{j=1}^{n-1} I_{mj}$

Subject to,

$$\sum_{j=1}^{n} x_{jk} = 1 \quad k = 1, \ldots, n$$

$$\sum_{k=1}^{n} x_{jk} = 1 \quad j = 1, \ldots, n$$

$$I_{ik} + \sum_{j=1}^{n} x_{j,k+1}p_{ij} + W_{i,k+1} - W_{ik} - \sum_{j=1}^{n} x_{jk}p_{i+1,j} - I_{i+1,k} = 0$$

$$for\ k = 1, \ldots, n-1; \quad i = 1, \ldots, m-1$$
$$W_{i1} = 0 \quad i = 1, \ldots, m-1$$
$$I_{1k} = 0 \quad k = 1, \ldots, n-1$$
$$x_{jk} \geq 0 \qquad integrs\ for \quad j, k = 1, \ldots, n$$
$$W_{ik}, I_{ik} \geq 0 \qquad\qquad i, k = 1, \ldots, n$$

The number of variables in this model is $n^2 + 2(m-1)(n-1)$ and the number of constraints is $nm + n - m + 1$. Flow shop problems with two or three stages can be solved using simple algorithms such as Johnson's method. Other methods such as branch and bound and dynamic programming are also available for solving larger size problems.

The model presented above can be extended for integrated scheduling in multiple production units in a supply chain with minor modifications to accommodate differences in objectives, priorities and conditions between independent units.

3.2.3 Modeling Job Shop Scheduling Problems

The job shop has a more general structure than the flow shop in which each job go through multiple processing stages (or machines) in an order that might be different than other jobs. The basic job shop scheduling problem holds the same assumptions that were made for the basic flow shop problem. In addition it is assumed that each job may be processed by a machine at most once, i.e., without recirculation.

Each job i has g_i operations and each operation j is processed in a different machine k. Thus each operation is identified by three indices, i, j, k: i, the job number to which the operation belongs, j is the operation's order in the job, and k is the machine processing that operation. Figure 3.6 gives an example of a job shop with 5 jobs each with multiple operations represented by blocks of a certain length proportional to their processing times. The requirement is to minimally arrange these 16 operations (blocks) into row of machines, without changing the length of any block or violating the basic assumptions of the job shop problem.

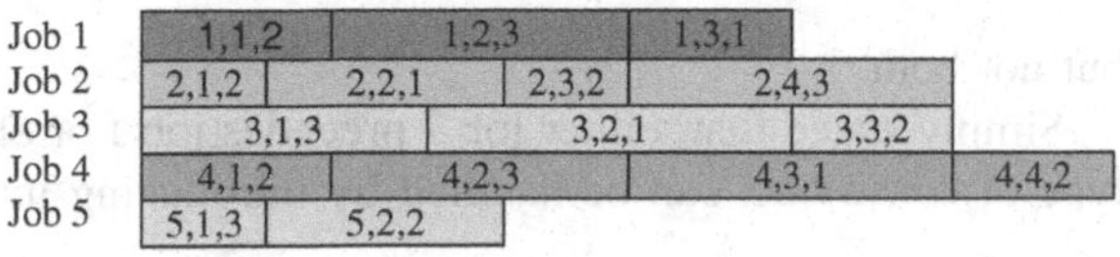

Fig. 3.5 Processing requirement of a 5 job 3 machine job shop problem

Job 1	1,1,2	1,2,3	1,3,1	
Job 2	2,1,2	2,2,1	2,3,2	2,4,3
Job 3	3,1,3	3,2,1	3,3,2	
Job 4	4,1,2	4,2,3	4,3,1	4,4,2
Job 5	5,1,3	5,2,2		

A colored Gantt chart of a possible schedule is shown in Fig. 3.7. Notice that no more than one operation of the same job is being processed at the same time and no machine is processing two operations at the same time. Also, the order of operations is maintained for each job across the schedule, as given in Fig. 3.6. Clearly this many possible schedules (block orders) satisfying the requirement, but it is needed to find the one with the shortest length (makespan).

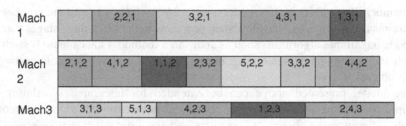

Fig. 3.6 A schedule for the problem in Fig. 3.5

This problem is extensively studied in the literature with various solution methods and several modelling approaches. One of the first Mixed Integer Programming models for the problem was developed by Manne [7] in 1960 and adopted here as described by Conway et al. [2].

3.2.3.1 MIP Model

Define the following variables.

$p_{ik} =$ the processing time of job I on machine k

$$r_{ijk} = \begin{cases} 1 & \text{if the jth operation of job i requires machine k} \\ 0 & \text{otherwise} \end{cases}$$

$T_{ik} =$ the starting time of job i on machine k

For the requirement that a machine can process only one job at a time, we have for two jobs, I and J, either,

$$(T_{Ik} - T_{Jk}) \geq p_{Jk} \text{ or } (T_{Jk} - T_{Ik}) \geq p_{Ik}$$

but not both.

Simply stated that, either job J precedes job I or else job I precedes job J. This type of constraints can be handled by introducing integer variables.

$$Y_{IJk} = \begin{cases} 1 & \text{if job } I \text{ precedes job } J \text{ (not necessarily directly) on machine } k \\ 0 & \text{otherwise} \end{cases}$$

The constraints can now be written as,

$$(M + p_{Jk})(Y_{IJk}) + (T_{Ik} - T_{Jk}) \geq p_{Jk}, \quad k = 1, \ldots, m$$
$$(M + p_{Ik})(1 - Y_{IJk}) + (T_{Jk} - T_{Ik}) \geq p_{Ik}, \quad k = 1, \ldots, m$$

The M is a constant and is chosen sufficiently large so that only one of the two inequalities is binding for $Y_{IJk} = 0$ or 1.

The operation precedence constraints are handled by noting that $\sum_{k=1}^{m} r_{ijk} T_{ik}$ is the starting time of the jth operation of job i. For all but the last operation of a job, one must have, a job must complete before the start of the next operation, i.e.,

$$\sum_{k=1}^{m} r_{ijk}(T_{ik} + p_{ik}) \leq r_{i,j+1,k} T_{ik}, \quad j = 1, \ldots, n, \; i = 1, \ldots, g_{i-1}$$

Thus, the model for minimizing the maximum flow time, $F_{\max}$ is,

Minimize $F_{\max}$

Subject to,

$$\sum_{k=1}^{m} r_{imk}(T_{ik} + p_{ik}) \leq F_{\max}, \quad i = 1, \ldots, n$$

$$\sum_{k=1}^{m} r_{ijk}(T_{ik} + p_{ik}) \leq r_{i,j+1,k} T_{ik}, \quad i = 1, \ldots, n, \quad j = 1, \ldots, g_{i-1}$$

$$(M + p_{Jk})(Y_{IJk}) + (T_{Ik} - T_{Jk}) \geq p_{Jk}, \quad \text{for each pair of jobs } I, J, k = 1, \ldots, m$$
$$(M + p_{Ik})(1 - Y_{IJk}) + (T_{Jk} - T_{Ik}) \geq p_{Ik}, \quad \text{for each pair of jobs } I, J, k = 1, \ldots, m$$

3.2.3.2 Disjunctive Programming Formulation

An alternative model for the job shop problem can be represented by a disjunctive graph. This model is constructed by Roy and Sussmann [10] and extracted here from Pinedo [8]. The job shop structure is modeled by a directed graph G with a set of N nodes and two sets of arcs A and B. Each operation (i, j) of job j on machine k is represented by a node in the graph. The A arcs, called *conjunctive* (solid), represent the routes of the jobs. If arc $(i, j) \rightarrow (k, j)$ is part of A, then job j has to be processed on machine i before proceeding to machine k, i.e., operation (i, j) precedes operation (k, j). Two operations that belong to two different jobs and that have to be processed on the same machine are connected to one another by

two B arcs, called *disjunctive* (broken) arcs that go in opposite directions. The
processing time of an operation (i, j) is represented by the length of the arc(s),
conjunctive as well as disjunctive, emanating from that node. In addition, there is a
source node U and a sink node V, which are dummy nodes. The source node U has
n conjunctive arcs emanating to the first operations of the n jobs and the sink node
V has n conjunctive arcs coming from all the last operations. The arcs emanating
from the source have length zero. Figure 3.8 gives a job shop example to illustrate
the graph, G = (N, A, B), mapping concept. This corresponding graph is shown in
Fig. 3.9.

Job 1	(1,1)		(2,1)		(3,1)
Job 2	(2,2)	(1,2)	(4,2)		(3,2)
Job 3	(1,3)	(2,3)		(4,3)	

Fig. 3.7 An example of a 3 job 4 machine job shop problem

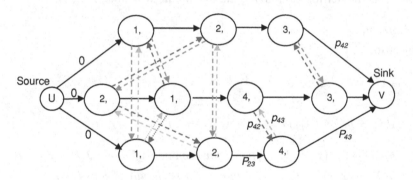

Fig. 3.8 Directed graph for job shop with makespan as objective

A feasible schedule corresponds to a *selection* of one disjunctive arc from each
pair such that the resulting directed graph is acyclic. The makespan of a feasible
schedule is determined by the longest path in $G(D)$ from the source U to the sink
V. This longest path consists of a set of operations of which the first starts at time 0
and the last finishes at the time of the makespan. Each operation on this path is
immediately followed by either the next operation on the same machine or the next
operation of the same job on another machine. The problem of minimizing the
makespan is reduced to finding a selection of disjunctive arcs that minimizes the
length of the longest path (that is, the critical path).

Therefore, the disjunctive programming formulation can be presented as
follows:

Let the variable y_{ij} denote the starting time of operation (i, j). Recall that set N denotes the set of all operations (i, j), and set A the set of all routing constraints $(i, j) \rightarrow (k, j)$ that require job j to be processed on machine I before it is processed on machine k. The following mathematical program minimizes the makespan.

Minimize $C_{\max}$

subject to,

$$y_{kj} - y_{ij} \geq p_{ij} \quad \text{for all } (i, j) \rightarrow (k, j) \in A$$

$$C_{\max} - y_{ij} \geq p_{ij} \quad \text{for all } (i, j) \in N$$

$$y_{ij} - y_{il} \geq p_{il} \text{ or } y_{il} - y_{ij} \geq p_{ij} \quad \text{for all } (i, l) \text{ and } (i, j), i = 1, \ldots, m$$

$$y_{ij} \geq 0 \quad \text{for all } (i, j) \in N$$

In this formulation, the first set of constraints ensure that operation (k, j) cannot start before completing operation (i, j). The third set of constraints ensures that some ordering exists among operations of different jobs that have to be processed on the same machine.

3.3 Supply Chain Scheduling

The structure of the supply chain is similar to a network of flow shops where independent units supply each other with raw material and services. However, each unit is a decision making unit that has its own objective, requirements, priorities and internal constraints. Each unit has its own planning and scheduling process starting at all levels from strategic, master scheduling, and operation scheduling. Coordination can be established at all levels. At a strategic level, decisions such as supplier selection, technology selection, and pricing may be coordinated across units, internal and external to the corporate. At the medium level (master planning), product types, capacities, distribution plans, and requirement planning are coordinated across plants. At the operational level, short term scheduling is optimized at each plant level and coordinated across plants and organization. The focus here is the scheduling level coordination in the supply chain. Each plant develops its own production schedule based on its objective and production structure. Job shops, flow shops, flexible job shops, bottleneck modelling are commonly used for optimally scheduling resources of short period of production time. Therefore, the models in the previous section are highly relative to supply chain scheduling coordination and integration. The simplest form of supply chain scheduling involves to units where the produced item in the supplier unit is required at a certain point of time at the customer manufacturing unit. Hence the completion time (plus transportation time) of the produced part is the release time of the product in the customer plant.

Fig. 3.9 A typical supply
chain network

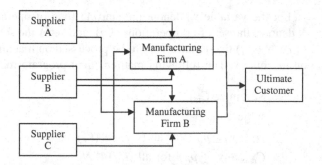

Consider for example a production unit supplied by one or more raw material
[5]. An order cannot be released before all the required has arrived. That is an
order j has an earliest possible starting time (release date r_j), a committed shipping
date d_j, and a priority factor or weight w_j. Every time a machine switches over
from one type of item to another, a setup S_{ijk} may be required and a setup cost may
be incurred. The supply chain model here is composed of a model within each
production unit depending on its structure, job shop, flowshop, single machine,
etc., and an objective function that links the two models. The objective to be
minimized may include the minimization of the total setup times and the total
tardiness T_j denoted as,

$$\alpha_1 \sum w_j T_j + \alpha_2 \sum I_{ijk} S_{ijk}$$

where α_1 and α_2 denote the weights of the two parts of the objective function. The
first part is the total weighted tardiness, and the second represents the total setup
times. I_{ijk} are indicator variables with value 1 if job j is followed by job k on
machine i, and 0 otherwise. This objective function links the two parts of the
supply chain through the objective function.

Another example of supply chain scheduling modeling is given by Luh and
Feng [6]. They consider a job shop manufacturing unit serving a series of manufacturers and suppliers (chain) each having his required quantities and internal
objective from his upstream supplier. Each supplier has his own internal production structure, constraints and objectives. The job shop scheduling model
within each unit is developed first using mixed integer programming formulation
with the objective of minimizing the expected total weighted earliness and tardiness (some of the problem parameters such as processing times are assumed to
be stochastic). The supply chain model is then constructed by adding cross-
member precedence constraints and summing up the individual objective functions
to form a global objective function. The problem is solved by Lagrangian relax-
ation method.

More general supply chain structures are studied in the literature. An example
of such structure is shown in Fig. 3.10. A model for a similar supply chain
structure will given next.

3.3.1 Integrated Medium Term Supply Chain Model

A model for a more general supply chain structure and requirement will be introduced to illustrate medium range integrated planning and scheduling in supply chains. The model is adopted from Kreipl and Pinedo [5].

Consider a supply chain of three levels in series shown in Fig. 3.10. The first and most upstream level (Level 1) has two factories in parallel producing two major products, F1 and F2, in full production capacity of 168 h (24 × 7) a week. They both feed a distribution center (DC) in Level 2 and deliver to a common customer in Level 3. Products can also be delivered to the customer by the distribution center. Both factories have no room for finished goods storage and the customer does not want to receive any early deliveries. The medium term planning production timing and quantities that minimizes the total cost of production cost, storage cost, transportation cost, tardiness cost for the whole supply chain over a 4 week time horizon. (the unit of time being one week). The transportation time from any one of the two factories to the DC, from any one of the two factories to the customer, and from the DC to the customer; all transportation times are assumed to be identical and equal to one week.

Fig. 3.10 A supply chain with three stages

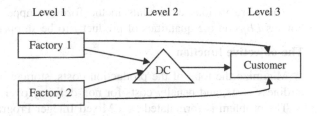

Parameters and inputs

D_{ijl} demand for product j, $j = 1, 2$, at levels l, $l = 2, 3$, by the end of week i, $i = 1, \ldots, 4$

tp_{jk} time (in hours) to produce 1,000 units of family j, $j = 1, 2$ in factory k, $k = 1, 2$

cp_{jk} unit cost of producing part j, $j = 1, 2$ *in factory k, $k = 1, 2$*

cs unit storage cost in DC of any product type

$c\tau 1_k$ unit cost of transportation from factory k, $k = 1, 2$, to DC

$c\tau 2_k$ unit cost of transportation from factory k, $k = 1, 2$, to the customer

$c\tau 3$ unit cost of transportation from DC to the customer

$t\tau$ transportation time between any two levels is assumed to be 1 week

TD_j tardiness penalty per unit per week of product j, $j = 1, 2$ that arrive late at the DC

TC_j tardiness penalty per unit per week of product j, $j = 1, 2$ that arrive late at the customer

TT penalty for never delivering one unit of product.

Decision variables

x_{ijk} number of units of product j produced at plant k during period i

y_{ijk2} number of units of product j transported from plant k to the DC in week i

y_{ijk3} number of units of product j transported from plant k to customer in week i

z_{ij} number of units of product j transported from the DC to the customer in week i

q_{0j2} number of units of product j in storage at the DC at time 0

q_{ij2} number of units of product j in storage at the DC in week i

v_{ij2} number of units of family j that are tardy (have not yet arrived) at the DC in week i

v_{4j2} number of units of product j that have not been delivered to the DC by the end of the planning horizon (the end of week 4)

v_{0j3} the number of units of product j that are tardy at the customer at time 0

v_{ij3} number of units of product j that are tardy at the customer in week i

v_{4j3} the number of units of product j that have not been delivered to the customer by the end of the planning horizon (the end of week 4).

Constraints

There are various constraints in the form of upper bounds UB_{jkl} and lower bounds LB_{jkl} on the quantities of product j to be shipped from plant k to stage l.

The objective function

Minimize the total of the production costs, storage costs, transportation costs, tardiness costs, and penalty costs for non-delivery over a horizon of 4 weeks.

The problem is formulated as a Mixed Integer Program as follows:

Minimize

$$\sum_{i=1}^{4}\sum_{j=1}^{2}\sum_{k=1}^{2}cp_{jk}x_{ijk} + \sum_{i=1}^{4}\sum_{j=1}^{2}\sum_{k=1}^{2}c\tau 1_{k}y_{ijk2} + \sum_{i=1}^{4}\sum_{j=1}^{2}\sum_{k=1}^{2}c\tau 2_{k}y_{ijk3} + \sum_{i=1}^{4}\sum_{j=1}^{2}c\tau 3z_{ij}$$

$$+ \sum_{i=1}^{4}\sum_{j=1}^{2}csq_{ij2} + \sum_{i=1}^{3}\sum_{j=1}^{2}TD_{j}v_{ij2} + \sum_{i=1}^{4}\sum_{j=1}^{2}TC_{j}v_{ij3}$$

$$+ \sum_{j=1}^{2}TTv_{4j2} + \sum_{j=1}^{2}TTv_{4j3}$$

Subject to,

Weekly production capacity constraints:

$$\sum_{j=1}^{2} tp_{j1}x_{ij1} \leq 168, \quad i = 1,\ldots,4;$$

$$\sum_{j=1}^{2} tp_{j2}x_{ij2} \leq 168, \quad i = 1,\ldots,4;$$

Transportation constraints:

For each, ijl, $i = 1,\ldots,4$, $j = 1, 2$, *and* $l = 2, 3$:

$$y_{ij1l} \leq UB_{j1l}$$
$$y_{ij1l} \geq UB_{j1l} \ or \ y_{ij1l} = 0$$
$$y_{ij2l} \leq UB_{j2l}$$
$$y_{ij2l} \geq UB_{j2l} \ or \ y_{ij2l} = 0$$

In addition to,

$$\sum_{l=2}^{3} y_{ijkl} = x_{ijk} \quad i = 1,\ldots,4, \quad j = 1,2, \quad k = 1,2;$$

$$\sum_{k=1}^{2} y_{ijk3} + z_{ij} \leq D_{i+1j,3} + v_{ij3} \quad i = 1,\ldots,3, \quad j = 1,2;$$

$$z_{1j} \leq \max(0, q_{0j2}) \quad j = 1,2;$$

$$z_{ij} \leq q_{i-1j,2} + y_{i-1j,1,2} + y_{i-1j,2,2} \quad i = 2,3,4, \quad j = 1,2;$$

Storage constraints:

$$q_{1j2} = \max(0, q_{0j2} - D_{1j2} - z_{1j}) \quad j = 1,2;$$
$$q_{ij2} = \max(0, q_{i-1j,2} + y_{i-1j,1,2} + y_{i-1j,2,2} - D_{ij2} - z_{ij} - v_{i-1j,2}) \quad i = 2,3,4 \ j = 1,2;$$

Constraints regarding number of jobs tardy and number of jobs not delivered:

$$v_{1j2} = \max(0, D_{1j2} - q_{0j2}) \quad j = 1,2;$$
$$v_{ij2} = \max(0, D_{ij2} + z_{ij} + v_{i-1j,2} - q_{ij,2} - y_{i-1j,1,2} - y_{i-1j,2,2}) \quad i = 2,3,4 \ j = 1,2;$$
$$v_{1j3} = \max(0, D_{1j3}) \quad j = 1,2;$$
$$v_{ij3} = \max(0, D_{ij3} + v_{i-1j,3} - z_{i-1j} - y_{i-1j,1,3} - y_{i-1j,2,3}) \quad i = 2,3,4 \ j = 1,2;$$

It is clear that most variables in this Mixed Integer Programming formulation are continuous variables.

Remarks

- As companies rely more on their business partners or suppliers, scheduling within production units has to be extended to coordinate between members in the supply chain.
- In this chapter, scheduling in supply chains is discussed in view of long time advances in machine scheduling theory. Classical models in flow shop and job shop scheduling are reviewed and introduced as a foundation for modeling supply chain scheduling. Models for 2-stage and 3-stage supply chain scheduling are then introduced to demonstrate modeling in supply chains.
- The models have different modeling approaches. One approach deals with independent models at each stage linked with the objective function and/or delivery requirement. The other is the integrated approach in which the whole chain is considered as a single production unit. Both approaches use traditional scheduling models as a base for modeling in supply chains.

References

1. Agnetis A, Hall NG, Pacciarelli D (2006) Supply chain scheduling: sequence coordination. Discrete Appl Math 154(15):2044–2063
2. Conway RW, Maxwell WL, Miller LW (2012) Theory of scheduling. Courier Dover Publications, New York
3. French S (1982) Sequencing and scheduling: an introduction to the mathematics of the job-shop, vol 683. Ellis Horwood, Chichester, p 684
4. Hall NG, Potts CN (2003) Supply chain scheduling: batching and delivery. Oper Res 51(4):566–584
5. Kreipl S, Pinedo M (2004) Planning and scheduling in supply chains: an overview of issues in practice. Prod Oper Manag 13(1):77–92
6. Luh PB, Feng W (2003) From manufacturing scheduling to supply chain coordination: the control of complexity and uncertainty. J Syst Sci Syst Eng 12(3):279–297
7. Manne AS (1960) On the job-shop scheduling problem. Oper Res 8(2):219–223
8. Pinedo M (2002) Scheduling: theory, algorithms, and systems, 2nd edn. Prentice Hall, New Jersey
9. Pinedo M (2005) Planning and scheduling in manufacturing and services, vol 24. Springer, New York
10. Roy B, Sussmann B (1964). Les problems d'ordonnancement avec contraintes disjonctives. Note ds, 9
11. Wagner HM (1995) An integer programming model for machine scheduling. Nav Res Logistics Q 6:131–140

Chapter 4
Optimization in Supply Chain

Abstract Transportation and facility location decisions are crucial in strategic supply chain design. Optimization models guide location decisions giving the optimal site selection under certain assumptions and constraints. It is an art to decide which model to use and how to modify the results based on the needs of a company. This chapter presents some of the important optimization models in supply chain. Mathematical formulations and solution procedures are also given. The models can be expanded for multi-echelon supply chains and/or include multiple products.

Keywords Facility location · Transportation · Linear programming · Integer programming

In Chap. 2, we dealt with topics in supply chain management. Supply chain management comprises decision making about facility location, production, transportation, and inventory control. Many companies employ optimization as a decision making tool. Here, we will introduce important and core optimization models and solution strategies for some important supply chain problems.

4.1 Transportation Problems

Transportation is flow of goods between supply chain stakeholders. The flow can be between and through any echelon of the supply chain: from warehouse to factory, from factory to customer etc. The transportation problem can be viewed as a network flow problem where the nodes represent stakeholders, edges represent the cost and amount of transportation between them basically. Consider the network in Fig. 4.1. S_n represents the amount of supply at node n. D_m is the amount of demand at node m. This network is a direct shipment network.

If the total supply is equal to the total demand, the problem is called balanced. If the problem is not balanced then dummy nodes (supply or demand) are

Fig. 4.1 Direct shipping
network

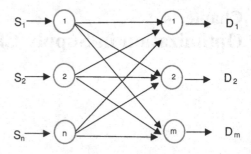

Fig. 4.2 Unit cost and
assignment amounts in a
transportation table

introduced. Supposing x_{ij} number of products to be shipped from S_i to D_j with a
unit cost of c_{ij}, we can write the mathematical model as follows:

$$\min \sum_{i=1}^{n} \sum_{j=1}^{m} c_{ij} x_{ij}$$

subject to

$$\sum_{j=1}^{m} x_{ij} = S_i, \quad i = 1, 2, \ldots, n$$

$$\sum_{i=1}^{n} x_{ij} = D_j, \quad j = 1, 2, \ldots, m$$

$$x_{ij} \geq 0, \quad \forall i, j$$

The problem has $m + n$ equations and $m \times n$ variables. First set of constraints
(equations) imply that total number of products shipped from a supplier to all of
the demand points is equal to the capacity of the supplier. The second set of
constraints imply that total number of shipments to a demand point is equal to the
size of the demand at the demand point. Based on the special structure of transportation problem, optimal solution can be obtained by:

1. Finding an initial solution (feasible)
2. Iterating over the initial solution to find an optimal solution.

Three widely used methods to finds a feasible initial solution are:

1. Northwest-corner method
2. Least-cost method
3. Vogel approximation method

It is intuitive to represent solutions using tables composed of cells such as the one
shown in Fig. 4.2.

Table 4.1 Transportation table

	D_1	D_2	D_3	D_4	Supply
S_1	5	2	3	6	15
S_2	10	6	7	8	25
S_3	5	4	7	9	15
Demand	10	15	10	20	

Table 4.2 Solution by northwest-corner method

	D_1	D_2	D_3	D_4	Supply
S_1	5 / 10	2 / 5	3	6	15
S_2	10	6 / 10	7 / 10	8 / 5	25
S_3	5	4	7	9 / 15	15
Demand	10	15	10	20	

Suppose that we are given three suppliers, four demand points transportation problem with the Table 4.1 including the unit transportation costs at the upper right corner of the cells of the table.

Northwest-corner method has three steps. Start with the upper left cell of the table. The corresponding variable is x_{11}.

1. Assign the selected cell as many products as possible. Update the supply and demand values based on the assignment amount.
2. Rows/columns with zero supply/demand are eliminated. If both of them are zero eliminate only one of them.
3. Stop if one row/column is left. Else, move to the right or below of the current cell. Go to the first step.

The solution found by this algorithm is shown in Table 4.2.

Table 4.3 Solution by the least-cost method

	D₁	D₂	D₃	D₄	Supply
S₁	5	2 15	3 0	6	15
S₂	10	6	7 5	8 20	25
S₃	5 10	4	7 5	9	15
Demand	10	15	10	20	

The cost associated with the solution is $5 \times 10 + 2 \times 5 + 6 \times 10 + 7 \times 10 + 8 \times 5 + 9 \times 15 = \$365$.

The least-cost method starts assigning products as many as possible to the least unit-cost cell. In case of a tie, one of the least-cost cell are chosen randomly. The row or column is eliminated and supply and demand are updated. In case of elimination of both a row and a column, one of them is eliminated. Then, remaining least unit-cost cell is assigned as many as possible. The procedure goes on till one row or column is left.

Working on the same example, we end up with the starting solution is shown on Table 4.3. The cost associated with the solution is $2 \times 15 + 3 \times 0 + 5 \times 10 + 7 \times 5 + 7 \times 5 + 8 \times 20 = \$310$.

Vogel approximation usually gives better initial solution. It is an improved version of the least-cost method [3]. The method has three steps:

1. A penalty is calculated finding the difference between two smallest unit-cost for each row and column.
2. The least unit-cost cell on the row or column with the largest penalty is assigned as many as possible. The row or column is eliminated. The supply and demand are updated. As in previous methods only the row or the column is eliminated in case of hitting zero supply and demand.
3. If only one row or column is left with 0 supply or demand, stop. If the only row or column has a positive supply or demand, basic variables of the row or column are determined based on the least-cost method. If the remaining rows and columns have zero supply and demand, zero basic variables are determined based on the least-cost method. Otherwise, return to step 1. Working on the same example we obtain the initial solution as shown in Tables 4.4, 4.5, 4.6, 4.7.

The associated cost is $3 \times 10 + 2 \times 5 + 4 \times 10 + 5 \times 5 + 10 \times 5 + 8 \times 20 = \$315$. Once we find an initial solution for the transportation problem we can

Table 4.4 First assignment of Vogel approximation

	D$_1$	D$_2$	D$_3$	D$_4$	Supply	R.Penalty
S$_1$	5	2	3 10	6	15	3-2=1
S$_2$	10	6	7	8	25	7-6=1
S$_3$	5	4	7	9	15	5-4=1
Demand	10	15	10	20		
C.Penalty	5-5=0	4-2=2	7-3=4	8-6=2		

Table 4.5 Penalty calculations after the first assignment

	D$_1$	D$_2$	D$_3$	D$_4$	Supply	R.Penalty
S$_1$	5	2	3 10	6	15	5-2=3
S$_2$	10	6	7	8	25	8-6=2
S$_3$	5	4	7	9	15	5-4=1
Demand	10	15	10	20		
C.Penalty	5-5=0	4-2=2	-	8-6=2		

develop the optimal solution for it using transportation simplex algorithm. The algorithm has two steps as follows:

1. The problem is checked based on simplex optimality condition. Entering variable is determined. If the solution is optimal, the algorithm stops. Otherwise, go to second step.
2. Leaving variable is determined based on the feasibility. Basis changes and return to first step.

The computations of the method are rooted in dual linear programming formulation and row operations similar to ones used in Simplex algorithm to solve linear programming problems. We can apply transportation algorithm using the initial solution from Vogel approximation method. As mentioned in Taha's book [3] multipliers u_i and v_j are associated with row i and column j of transportation table.

Table 4.6 Second and third assignments of Vogel approximation

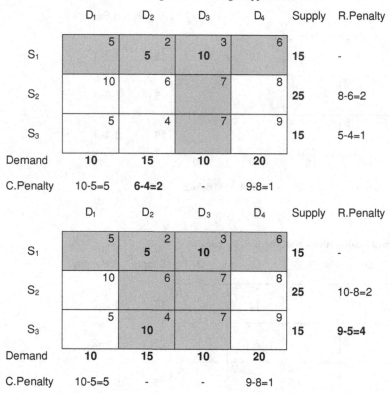

	D₁	D₂	D₃	D₄	Supply	R.Penalty
S₁	5	2 **5**	3 **10**	6	15	-
S₂	10	6	7	8	25	8-6=2
S₃	5	4	7	9	15	5-4=1
Demand	**10**	**15**	**10**	**20**		
C.Penalty	10-5=5	6-4=2	-	9-8=1		

	D₁	D₂	D₃	D₄	Supply	R.Penalty
S₁	5	2 **5**	3 **10**	6	15	-
S₂	10	6	7	8	25	10-8=2
S₃	5	4 **10**	7	9	15	**9-5=4**
Demand	**10**	**15**	**10**	**20**		
C.Penalty	10-5=5	-	-	9-8=1		

Table 4.7 Solution by Vogel approximation

	D₁	D₂	D₃	D₄	Supply
S₁	5	2 **5**	3 **10**	6	15
S₂	10 **5**	6	7	8 **20**	25
S₃	5 **5**	4 **10**	7	9	15
Demand	**10**	**15**	**10**	**20**	

For each basic variable x_{ij} (the variables on assigned cells of the transportation table) on the table the following relation holds:

$$u_i + v_j = c_{ij}$$

Starting solution on table has six basic variables (that means six equations) and seven unknowns. Once we set one of the multiplier values equal to zero, we can find a unique solution. For example, setting v_1 to 0, we can find the remaining unknowns as follows:

$$u_2 + v_1 = 10(c_{21} \text{ value associated with the assignment } x_{21} = 5)$$

We can show the equations and solutions with associated basic variables in the following manner: basic variable, equation, and multiplier value.

$$x_{21}, u_2 + v_1 = 10, \; u_2 = 10$$
$$x_{31}, u_3 + v_1 = 5, \; u_3 = 5$$
$$x_{32}, u_3 + v_2 = 4, \; v_2 = -1$$
$$x_{12}, u_1 + v_2 = 2, \; u_1 = 3$$
$$x_{13}, u_1 + v_3 = 3, \; v_3 = 0$$
$$x_{24}, u_2 + v_4 = 8, \; v_4 = -2$$

Now, non-basic variables (candidates to be assigned some amount of shipment or to be an entering variable) are evaluated checking the reduced cost values:

$$u_i + v_j - c_{ij}$$

We can show the calculations as: non-basic variable, reduced cost.

$$x_{11}, 3 + 0 - 5 = -2$$
$$x_{14}, 3 - 2 - 6 = -5$$
$$\mathbf{x_{22}, 10 - 1 - 6 = 3}$$
$$x_{23}, 10 + 0 - 7 = 3$$
$$x_{33}, 5 + 0 - 7 = -2$$
$$x_{34}, 5 - 2 - 9 = -6$$

We chose the highest positive value and corresponding non-basic variable since the objective is minimization. That means the entering variable will improve the objective function more than the others (among positive reduce costs). We arbitrarily choose x_{22} among the basic variables since value of three is associated with two of them. One basic variable should leave. Amount of Q (as many products as possible) is assigned to x_{22}. The maximum amount to be assigned is subject to supplier capacity constraints and demand point constraints. Using the transportation table, a closed loop starting from the entering variable cell and ending at the same cell is constructed. The loop is constructed using connected horizontal and vertical lines. Corner points excluding the entering variable must be on a basic variable. The loop is shown on Table 4.8. Assigning Q to x_{22} must satisfy the following constraints:

Table 4.8 Closed loop for assigning Q

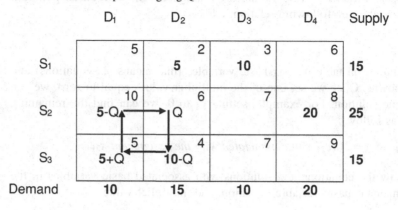

	D$_1$	D$_2$	D$_3$	D$_4$	Supply
S$_1$	5	2 5	3 10	6	15
S$_2$	10 5-Q	6 Q	7	8 20	25
S$_3$	5 5+Q	4 10-Q	7	9	15
Demand	10	15	10	20	

Table 4.9 New solution after the first iteration

	D$_1$	D$_2$	D$_3$	D$_4$	Supply
S$_1$	5	2 5	3 10	6	15
S$_2$	10	6 5	7	8 20	25
S$_3$	5 10	4 5	7	9	15
Demand	10	15	10	20	

$$10 - Q \geq 0$$
$$5 - Q \geq 0$$

The maximum assignment can be five products. At the end of the first iteration, the new solution is shown in Table 4.9.

The algorithm is repeated based on the new solution. It turns out that first iteration gives the optimal solution (reduced costs based on the new solution are non-positive). The objective function is improved by reduced cost × amount of assignment, $3 \times 5 = 15$. The value of the objective function is $315 - 15 = \$300$.

The coefficient matrix of the transportation problem is unimodular. This feature guarantees integer solution for a linear programming problem.

We can incorporate warehouses between suppliers and demand points, and transship through the warehouses. In other words, warehouses can both receive shipments and send shipments (behaving like both supplier and demand point). Then, the problem is called transshipment problem. The transshipment problem is a special case of minimum cost capacitated network model where capacity constraints are removed.

In addition to classical transportation network, transshipment network, other network designs for transportation exist. *Milk run* [1] route is one of them. Direct shipping with milk runs includes deliveries from a single supplier to multiple demand points or from multiple suppliers to a demand point. Here, the idea is loading a truck and distributing the products on a route where demand points (or suppliers) lay or collecting deliveries on a route of suppliers to deliver to a demand point. Determination of the route for each milk run here can be approached by heuristic methods or variants of a travelling salesman problem (TSP). Milk run design can lower transportation cost by loading the truck more compared to classical transportation as long as demand points (or suppliers) are close to each other.

Warehouses can hold products to achieve economies of scale. If a company's customers are accumulated close to each other inside a geographic region, the company might deliver massive shipments to its warehouse close to customers to decrease the inbound logistic costs. Then products may be cross docked or shipped to customers in smaller lots with the potential to save from outbound logistic costs as well. Such a supply chain design decision is made after trade-off calculations for cost of opening a warehouse, holding costs, and logistic costs. There might be other transportation network designs based on the need and constraints of the supply chain.

Remarks

- Transportation problems can be viewed as network flow problems.
- In balance transportation problems, total supply is equal to total demand.
- Classical transportation problem can be solved optimally finding a feasible solution using north-west, least-cost or Vogel approximation method and then using transportation simplex algorithm.
- Transshipment problem includes intermediate nodes that behave as both a supplier and a demand point.
- There may be different transportation network designs within a supply chain such.
- Transportation network design decisions are affected by the cost of opening new facilities, holding inventory costs, inbound, outbound logistic costs among others.

4.2 Facility Location Problems

The models and arguments in this section are mostly based on Watson et al. (2013)'s book [4]. Location problem are very diverse. American Mathematical Society (AMS) has specific codes for location problems (90B80 for discrete location and assignment, and 90B85 for continuous location) [2]. General location problems include customers and facilities to satisfy customer demands. Facility locations problems are classified as discrete and continuous ones. Here, we are interested in discrete facility location problems. Also problem distinction is based on being capacitated or not. Melo et al. [2] identify four core features to be included in a facility location model to use in supply chain decisions:

1. Multi-layer facilities
2. Multiple products (integration of bills of materials into supply chain design received the attention of many researchers)
3. Single or multiple periods (about 82 % of the papers they surveyed include single-period problems)
4. Deterministic or stochastic parameters

They reveal that facility location problems mostly include inventory and production decisions as well while routing, transportation mode selection, and procurement integrated location decision problems receive less attention in the literature. Facility location decisions are strategic in supply chain design since a company supply chain will need to adapt to changing market needs migrating to new locations for example. They also state that most of the facility location studies consider minimization of costs as the objective.

A simple location model is borrowed from physics that is called gravitation model. If want to locate a new facility with coordinates of (x, y) around n demand points with coordinates of (x_i, y_i), the (x, y) values are found by the following equations:

$$x = \frac{\sum_{i=1}^{n} x_i D_i}{\sum_{i=1}^{n} D_i}, \quad y = \frac{\sum_{i=1}^{n} y_i D_i}{\sum_{i=1}^{n} D_i}$$

These values are obtained based on minimizing the squared distances d_i from new facility to customer i times customer demand D_i values:

$$\min \sum_{i=1}^{n} d_i^2 D_i$$

For example, Table 4.10 gives the coordinates and demands of customers 1, 2, and 3. We find the location (x, y) of a new supply facility. It is like finding the gravity center of different masses in physics.

Table 4.10 Location data and solution

Customer	x-coordinate	y-coordinate	Demand
1	50	60	100
2	30	20	80
3	60	40	90
4	20	30	120
5	10	10	130
6	25	5	70
Supply facility location		x	y
		30.93	27.88

Thick circle in Fig. 4.3 shows the location of the new facility.

This model reflects a physics perspective, however this model might not fit a regular supply chain design since objective function employs a second degree penalty for unit distance travelled. However, gravity location model can give insight for potential location areas. Every model is an abstraction of reality that comes with assumptions. This model is no exception. This model does not take into account the physical features of location areas, i.e. mountainous or not, proximity to labor force or required infrastructure etc.

Another model for single facility layout problem has an objective of minimizing the summation of weighted distances between the new facility and existing ones (also known as minimum facility location problems, also minimax problems exist in the literature):

$$\min \sum_{i=1}^{n} D_i d_i$$

Here, weights are demands of customers. Weights can also be number of trips, shipments etc. A special case of this problem is when rectilinear distance (known as L_1 norm) values are used. The rectilinear distance between the new facility and a demand point is found by:

$$|x - x_i| + |y - y_i|$$

The following algorithm find the (x, y) values that will give the optimal solution to this problem. The reasoning behind the algorithm is the fact that the objective function can be represented as two separate functions. Each function is a convex function that is guaranteed to have a global minimum point.

1. Sort x_i (y_i) values in ascending order
2. Find the cumulative weights for each value
3. x is the value at which cumulative weight value is greater than or equal to the half of the total weight.

Working on the same example above (x, y) values are found to be (25, 30). Computations are shown in Tables 4.11 and 4.12:

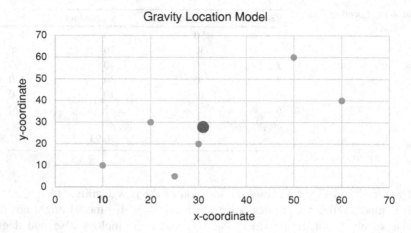

Fig. 4.3 Gravity location model

The same problem can be formulated as a linear program (LP). General LP formulation is given below:

$$\min \sum_{i=1}^{n} w_i(r_i + s_i)$$
$$subject\ to$$
$$x - r_i + s_i = x_i; \quad \forall i$$
$$r_i, s_i \geq 0, \quad \forall i$$

For our example the LP has twelve variables and six constraints. The optimal solution can be found using simplex algorithm or common software such as excel solver employing similar algorithms.

w_i	w_i value	x_i	x_i value	r_i	r_i value	s_i	s_i value	r_i+s_i	Constraints	LHS	RHS
w_1	100	x_1	50	r_1	0	s_1	25	25	1	50	50
w_2	80	x_2	30	r_2	0	s_2	5	5	2	30	30
w_3	90	x_3	60	r_3	0	s_3	35	35	3	60	60
w_4	120	x_4	20	r_4	5	s_4	0	5	4	20	20
w_5	130	x_5	10	r_5	15	s_5	0	15	5	10	10
w_6	70	x_6	25	r_6	0	s_6	0	0	6	25	25
		x			25		Obj			8600	

Table 4.12 contains the required input data for excel solver. Lighter shaded cells include variable values. Darker shaded cells include excel formulations. The dark shaded cell corresponding to objective function (Obj) has the excel expression $= sumproduct\ (column(r_i + s_i),\ column(w_i))$. Constraint 1 LHS (left hand

Table 4.11 Customers (demand points) are sorted in ascending order of x-coordinates

Customer	x-coordinate	y-coordinate	Demand	Cumulative
5	10	10	130	130
4	20	30	120	250
6	25	5	70	320
2	30	20	80	400
1	50	60	100	500
3	60	40	90	590

x-coordinate value for which the cumulative demand value is greater than or equal to 295 (half of the total demands) for the first time is 25

Table 4.12 Customers (demand points) are sorted in ascending order of y-coordinates

Customer	x-coordinate	y-coordinate	Demand	Cumulative
6	25	5	70	70
5	10	10	130	200
2	30	20	80	280
4	20	30	120	400
3	60	40	90	490
1	50	60	100	590

y-coordinate value for which the cumulative demand value is greater than or equal to 295 (half of the total demands) for the first time is 30

side) dark shaded cell has the excel expression $= x\ cell - r_1\ value\ cell + s_1\ value\ cell$. Remaining dark shaded cells has similar expressions. Then, we click on solver button inside data tab (you might need to install solver add-in). Figure 4.4 shows the excel solver dialogue box. Please note that before hitting the solve button of excel solver, all shaded cells were empty.

Set objective cell has the address of *Obj* cell (address is passed as we click on it from the table). *Min* is chosen since our objective is minimization. *By Changing Variable Cells* cell has the variables (r_i value and s_i value column cells and x value cell are selected from the table). Constraints are added using add button. As we click on add button we are asked to enter left hand side values, logical operator ($\leq$, $\geq$, $=$, etc.) and right hand side values. We select LHS column cells, greater than or equal to operator, and RHS column cell values. We make sure that Make Unconstrained Variables Non-Negative box is checked. Then we click on solve. As we click, shaded cells are filled. We see that x value is 25. The same procedure is repeated to find y value that is 30. The following integer programming models are solved in a similar way using excel solver except that integer constraints are added (using *int* logical operator).

Location decisions are associated with transportation and/or production related decisions. For example, one might be interested in choosing a number of facility locations among alternatives. In other words, one might be subject to opening *F* number of facilities out of *N* to minimize total weighted distance from *F* facilities to stores satisfying all *C* customer demands and assuming that each facility

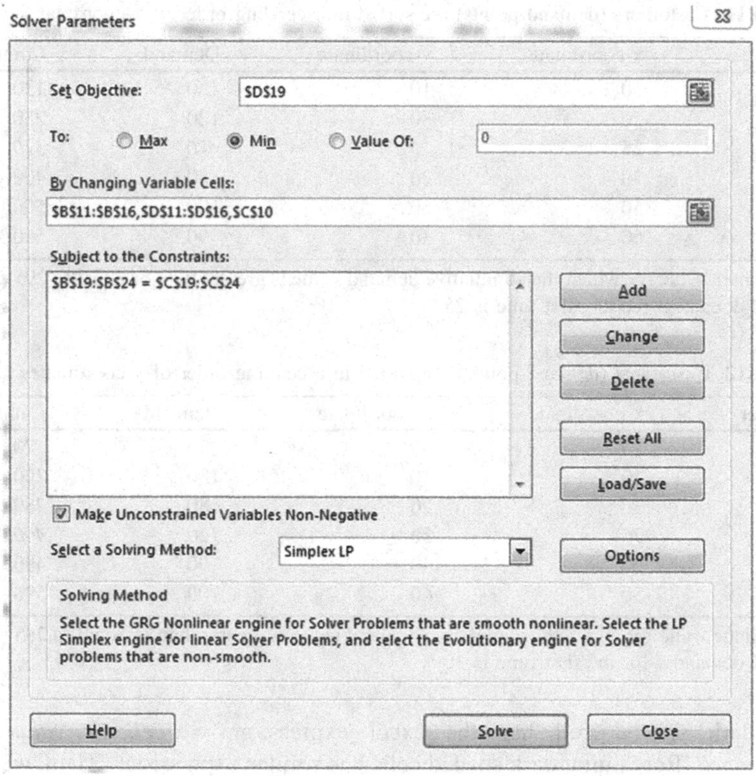

Fig. 4.4 Excel solver dialogue box

can satisfy customer demands. The following is an integer programming model for this type of problems (also called *P-median* problem or k-median clustering problem both of which are studied very well in the literature):

$$\min \sum_{i=1}^{F} \sum_{j=1}^{C} d_{i,j} D_j Y_{i,j}$$
$$subject\ to$$
$$\sum_{i=1}^{F} Y_{i,j} = 1; \quad \forall j$$
$$\sum_{i=1}^{F} X_i = F$$
$$Y_{i,j} \leq X_i; \quad \forall i,\ \forall j$$
$$Y_{i,j} \in \{0,1\}; \quad \forall i,\ \forall j$$
$$X_i \in \{0,1\}; \quad \forall i$$

This model suggests selection of F new facilities using X_i binary variables. X_i will be equal to one if the facility on site i is selected. $Y_{i,j}$ variables are introduced to choose which facility to serve (manufacturing plant ship to retailer) which customer. $Y_{i,j}$ will be equal to one if the new facility at location i servers customer j. The first set of constraints guarantee that each customer demand is satisfied by one of the new facilities. The second constraint implies that total number of new facilities opened is F. The third set of constraints indicate that a shipment from a new facility cannot be realized unless this new facility is opened. Remaining are binary constraints.

IP problems are hard problems (technically called NP-hard, 0-1 IP problems are NP-complete). The solution time for the problem grows enormously with the size of the problem (number of variables, number of constraints). One can imagine the intractability in enumerating all solutions and picking the best one as the optimal solution. Excel solver can be used to solve small instances of the problem.

The model can be extended considering capacity (labor, equipment etc.) constraints. Let's assume that each new facility has a capacity of K_i and $V_{i,j}$ is the volume of demand j satisfied by facility i. The only difference to the model above is the set of constraints:

$$\sum_{j=1}^{C} V_{i,j} Y_{i,j} \leq K_i X_i; \quad \forall i$$

By these constraints, demands satisfied from a new facility cannot exceed the capacity of it.

Instead of minimizing the total weighted distance, we can set our objective as minimizing the total weighted costs associated with shipping from facility i to customer j. Defining c_{ij} as the cost of shipping one unit of demand from facility i to customer j. The objective function becomes:

$$\min \sum_{i=1}^{F} \sum_{j=1}^{C} c_{i,j} D_j Y_{i,j}$$

We may have size options for the new facilities and associated operating costs. These costs are fixed ($f_{i,o}$) and variable costs (v_i) that can be reflected on the model. Hence, our selection variables are modified as $x_{i,o}$. The value of the variable is equal to one if facility on site i is to be opened with size option o. Variable costs are added to transportation costs. So, objective function becomes:

$$\sum_{i=1}^{F} \sum_{j=1}^{C} (c_{i,j} + v_i) D_j Y_{i,j} + \sum_{i=1}^{F} \sum_{o=1}^{O} f_{i,o} X_{i,o}$$

In order to prevent choosing more than one option for a new facility the following constraint is added:

$$\sum_{o=1}^{O} X_{i,o} \leq 1; \quad \forall i$$

We can modify capacity constraints to include size options such that:

$$\sum_{j=1}^{C} V_{i,j} Y_{i,j} \leq \sum_{0=1}^{O} K_{i,o} X_{i,o}; \quad \forall i$$

Also, we cannot send any item from a new facility to customers unless the new facility with size option is opened:

$$Y_{i,j} \leq \sum_{o=1}^{O} X_{i,o}; \quad \forall i, \forall j$$

If we want the number of facilities to open between two numbers that is $F_{\min}$, $F_{\max}$ we can include the following constraints in the model:

$$\sum_{i=1}^{F} \sum_{o=1}^{O} X_{i,o} \geq F_{\min}$$

$$\sum_{i=1}^{F} \sum_{o=1}^{O} X_{i,o} \leq F_{\max}$$

New selection variables should be binary as well:

$$X_{i,o} \in \{0,1\}; \quad \forall i, \forall o$$

We can expand our model for multi-echelon supply chains. Instead of dealing with facility customer relations, we can observe manufacturing facility, warehouse, and customer (i.e. retailer store) relationships. Here we need to re-define our variables, parameters and introduce new variables. Let tMW be transportation cost from manufacturing facility to warehouse, tWC transportation cost from warehouse to customer, Wv warehouse variable cost, Wf warehouse fixed cost, Mv manufacturing facility variable cost, MK manufacturing facility capacity, WK warehouse capacity. New variable $Z_{s,i}$ is the amount of shipments from manufacturing facility s to warehouse i.

We can build our new model, which is similar to the previous developed one, using new definitions:

$$\min \sum_{s=1}^{S} \sum_{i=1}^{F} \left(tMW_{s,i} + Mv_s\right)Z_{s,i} + \sum_{i=1}^{F} \sum_{j=1}^{C} \left(tWC_{i,j} + Wv_i\right)D_jY_{i,j} + \sum_{i=1}^{F} \sum_{o=1}^{O} Wf_{i,o}X_{i,o}$$

$$subject\ to$$

$$\sum_{i=1}^{F} Y_{i,j} = 1; \quad \forall j$$

$$\sum_{i=1}^{F} \sum_{o=1}^{O} X_{i,o} \geq F_{\min}$$

$$\sum_{i=1}^{F} \sum_{o=1}^{O} X_{i,o} \leq F_{\max}$$

$$\sum_{o=1}^{O} X_{i,o} \leq 1; \quad \forall i$$

$$\sum_{j=1}^{C} V_{i,j}Y_{i,j} \leq \sum_{0=1}^{O} WK_{i,o}X_{i,o}; \quad \forall i$$

$$\sum_{s=1}^{S} Z_{s,i} = \sum_{j=1}^{C} D_jY_{i,j}; \quad \forall i$$

$$\sum_{i=1}^{F} Z_{s,i} \leq MK_s; \quad \forall s$$

$$Y_{i,j} \leq \sum_{o=1}^{O} X_{i,o}; \quad \forall i, \forall j$$

$$Y_{i,j} \in \{0,1\}; \quad \forall i, \forall j$$

$$X_{i,o} \in \{0,1\}; \quad \forall i, \forall o$$

$$Z_{s,i} \geq 0; \quad \forall s, \forall i$$

Sixth constraint is called 'conservation of flow'. Left hand side of the equation imply the number of items coming into a warehouse while right hand side imply the number of items going out of the same warehouse. So, warehouses are neither consumers nor producers. Seventh constraint is capacity constraint for manufacturing facility.

The models mentioned till here assume single product distribution through the supply chain. We can include multiple products with minor change in the model (adding a product index to relevant variables, parameters). Demands and costs are deterministic. Probabilistic models exist. Planning horizon is single period. As parameters change over time multi-period modeling can be adopted. These changes make models complicated to solve efficiently. There is a trade-off between a model's being close to reality and solvability. The more complicated the problem, the harder to solve it.

Remarks

- Facility location decisions are crucial in strategic supply chain design. Optimization models guide location decisions giving the optimal site selection under certain assumptions and constraints. It is an art to decide which model to use and how to modify the results based on the needs of a company.

- Gravity model is easy to solve and a good start for location decision evaluation. However, it does not fit supply chain context very well.
- Locating a new facility to minimize the total weighted distance from the facility to customers is an easy to solve problem assuming rectilinear distance values. This problem can be modeled as a linear programming problem. Linear programming (LP) problems are easy and efficient algorithms exist that solve them optimally.
- Excel solver can be used to solve LP models and also some integer programming (IP) problems.
- Both LP and IP are optimization tools used especially by operations research people. These tools find optimal solutions to problems that include an objective and some constraints. LP variables are defined in positive real numbers domain while IP variables are defined in positive integer numbers domain. IP problems are hard to solve. The models reviewed in this chapter are linear integer programming model (ILP). If a model employs both continuous and integer variables, it is called mixed integer programming (MIP).
- P-median or k-median clustering problems are well studied in the literature. The IP model has the objective of minimizing total weighted distance from F facilities to stores satisfying all C customer demands and assuming that each facility can satisfy customer demands. The model can be extended to have capacity constraints. Minimizing total weighted costs can be employed as an objective function.
- We can extend P-median model with cost minimization to incorporate selection of new facilities with different size options and assigning fixed and variables costs.
- The models can be expanded for multi-echelon supply chains and/or include multiple products.

References

1. Chopra S, Meindl P (2004) Supply chain management, 2nd edn
2. Melo MT, Nickel S, Saldanha-da-Gama F (2009) Facility location and supply chain management—a review. Eur J Oper Res 196:401–412
3. Taha HA (2011) Operations research. Pearson, New Jersey
4. Watson M, Lewis S, Cacioppi P, Jayaraman J (2013) Supply chain network design: applying optimization and analytics to the global supply chain. FT Press, New Jersey

Printed in the United States
By Bookmasters